Viktoria Blüschke-Nikolaeva

OPTCON2

Viktoria Blüschke-Nikolaeva

OPTCON2

An Algorithm for the Optimal Control of Nonlinear Stochastic Models

Südwestdeutscher Verlag für Hochschulschriften

Impressum / Imprint

Bibliografische Information der Deutschen Nationalbibliothek: Die Deutsche Nationalbibliothek verzeichnet diese Publikation in der Deutschen Nationalbibliografie; detaillierte bibliografische Daten sind im Internet über http://dnb.d-nb.de abrufbar.

Alle in diesem Buch genannten Marken und Produktnamen unterliegen warenzeichen-, marken- oder patentrechtlichem Schutz bzw. sind Warenzeichen oder eingetragene Warenzeichen der jeweiligen Inhaber. Die Wiedergabe von Marken, Produktnamen, Gebrauchsnamen, Handelsnamen, Warenbezeichnungen u.s.w. in diesem Werk berechtigt auch ohne besondere Kennzeichnung nicht zu der Annahme, dass solche Namen im Sinne der Warenzeichen- und Markenschutzgesetzgebung als frei zu betrachten wären und daher von jedermann benutzt werden dürften.

Bibliographic information published by the Deutsche Nationalbibliothek: The Deutsche Nationalbibliothek lists this publication in the Deutsche Nationalbibliografie; detailed bibliographic data are available in the Internet at http://dnb.d-nb.de.

Any brand names and product names mentioned in this book are subject to trademark, brand or patent protection and are trademarks or registered trademarks of their respective holders. The use of brand names, product names, common names, trade names, product descriptions etc. even without a particular marking in this works is in no way to be construed to mean that such names may be regarded as unrestricted in respect of trademark and brand protection legislation and could thus be used by anyone.

Coverbild / Cover image: www.ingimage.com

Verlag / Publisher:
Südwestdeutscher Verlag für Hochschulschriften
ist ein Imprint der / is a trademark of
AV Akademikerverlag GmbH & Co. KG
Heinrich-Böcking-Str. 6-8, 66121 Saarbrücken, Deutschland / Germany
Email: info@svh-verlag.de

Herstellung: siehe letzte Seite /
Printed at: see last page
ISBN: 978-3-8381-3636-3

Zugl. / Approved by: Klagenfurt, Uni, Diss., 2012

Copyright © 2013 AV Akademikerverlag GmbH & Co. KG
Alle Rechte vorbehalten. / All rights reserved. Saarbrücken 2013

Contents

1 Introduction 3

2 Basic terms and concept 5
   2.1 The optimum control problem 5
   2.2 Literature survey 6

3 The OPTCON1 algorithm 8
   3.1 Elements of OPTCON1 9
   3.2 Schematic description of OPTCON1 15

4 The OPTCON2 algorithm 19
   4.1 Elements of OPTCON2 19
   4.2 Schematic description of OPTCON2 28

5 Applications 31
   5.1 The SLOVNL model 32
   5.2 The SLOVL model 38
   5.3 The MacRae model 43
   5.4 The Abel model 45

6 Conclusion 48

7 Appendix 49
   7.1 Theorem 1 49
   7.2 Theorem 2 50
   7.3 Software description 52
   7.4 Diagrams 55

8 Acknowledgements 59

References 60

CONTENTS

1 Introduction

In particular, there have been many studies determining optimal policies for management and economic models. In these cases, the planner or decision-maker wants a firm or an economy to behave in such a way as to obtain time paths for some variables of interest that are 'optimal' in some well defined sense, i.e. they maximize or minimize some objective function. For example, the manager of a firm may want to maximize profits or minimize costs subject to the restrictions posed by the firm's current production technology or available personnel, etc. Or a policy-maker responsible for fiscal policy in a country may determine tax rates and government expenditures so as to obtain the most desired ('ideal') time paths for such objective variables as the rate of growth of real GDP, the rate of unemployment or the rate of inflation, among others. This is a very difficult problem to study, particularly when the dynamic system is nonlinear and stochastic. Most optimum control applications use algorithms for linear dynamic systems or those that do not take the full stochastic nature of the econometric model into account. Examples of the former are Kendrick (1981), Coomes (1987) and the references in Amman (1996), and Chow (1975, 1981) for the latter. An algorithm that is explicitly aimed at providing (approximate) solutions to optimum control problems for nonlinear econometric models with additive and multiplicative uncertainties is OPTCON, as introduced by Matulka and Neck (1992). However, so far OPTCON has been severely limited by being based on very restrictive assumptions about the information available to the decision-maker. In particular, OPTCON does not include any kind of learning about the econometric model while in the process of controlling the economy. In reality, however, new information arrives in each time period, and econometric models are regularly re-estimated using this information. Motivated by a desire to reduce OPTCON's learning deficiency, the research in this direction is accomplished and the results are presented in this book.

The development from open-loop control only (the basic OPTCON algorithm, also named OPTCON1) to the inclusion of passive learning or open-loop feedback control, where the estimates of the parameters are updated in each time period, results in the OPTCON2 algorithm. Thus, the new version of the algorithm can deliver numerical approximate solutions to optimum control problems for a large class of nonlinear dynamic systems under a quadratic objective function with stochastic uncertainty in the parameters and in the system equations under both kinds of control schemes. In the open-loop case the solution of the model is computed for all time periods at once without observing current information. In contrast, in the open-loop feedback case it is assumed that the new realizations of the two random processes mentioned above occur in each period, which can then be used to re-estimate the uncertain parameters of the dynamic system, i.e. the parameters of the econometric model. Following Kendrick's (1981) approach, the parameter estimates are updated using the Kalman filter in expectation of a more reliable approximation to the solution of the stochastic optimum control problem.

In the new version of the algorithm the theoretical developments, their corresponding implementation in the computer language *C#* and several applications of the OPTCON2 algorithm to different macroeconomic problems are carried out.[1]

The aim of the application part is threefold. Firstly, the convergence and applicability of the OPTCON2 algorithm for solving optimum control problems should be shown. For this purpose two existing models are used, MacRae and Abel, as well as two new models, SLOVNL and SLOVL, which were developed for the purposes of this research. The second research question to be answered with the application concerns the 'correct' functionality of the OPTCON2 algorithm. Whether the passive learning control strategy in OPTCON2 gives 'correct' results is checked by comparing the results of OPTCON2 with the solutions of Kendrick's algorithm DUAL where a passive learning strategy is included. To this end the MacRae and

---

[1] The implementation of the algorithm in *C#* was done in cooperation with D. Blueschke and the application part was mainly done in cooperation with R. Neck and D. Blueschke.

1 INTRODUCTION

Abel models are applied. The third aim is to check whether passive learning can improve the final results or not. For this purpose the OPTCON2 algorithm is applied to all four models. Moreover, an additional research question is analyzed by applying the OPTCON2 algorithm to the SLOVNL and SLOVL models, namely the influence of the stochastic on the final results. In addition the impact of the weights within the updating procedure on the optimal solutions is studied using the SLOVNL model.

This book has the following structure: in Section 2, the basic terms and the concept are given. Section 3 reviews the basic version of the algorithm, whereby the algorithmic schema and characteristics as well as the elements of the OPTCON1 algorithm are discussed. Section 4 explains the OPTCON2 algorithm; in particular all the innovations that have been included in the basic version of the algorithm are discussed in precise detail. In Section 5 the application of the OPTCON2 algorithm to four macroeconomic models is presented, namely two well known models, MacRae and Abel, and two new models of the Slovenian economy, SLOVNL and SLOVL. The dokument is rounded off in Section 6 with a summary of the results and a set of conclusions that can be drawn.

2 Basic terms and concept

In this section the class of problems to be tackled by the OPTCON1 and the OPTCON2 algorithms as well as some examples for such kind of problems are presented. In addition an overview of the literature is given.

2.1 The optimum control problem

First, the general form of an optimum control problem has to be defined.
The general dynamic optimum control problem in discrete time consists of the criterion or objective function J which is to minimize (or to maximize)

$$\min_u J = \sum_{t=0}^{T} L(x_t, u_t) \qquad (1)$$

and an appropriate dynamic in the form of a system of equations[2]

$$x_t = f(x_{t-1}, x_t, u_t, z_t), \quad t = 1, ..., T; \qquad (2)$$
$$x_0 \text{ is given.}$$

There are two major kinds of variables in this problem, namely state variables x_t that describe the system at the time point t and the control variables or the policies u_t that can be controlled during the process. z_t denotes a vector of non-controlled exogenous variables. In the context of macroeconomics the control variables (e.g. tax, government expenditure, interest rates) are directly manipulated or controlled by the government authorities or central bank in order to achieve the desired levels of the state variables (e.g. unemployment, GDP, inflation). In a microeconomic problem, where the profit has to be maximized for example, the turnover can be used as a state variable and the price as a control variable which the firm can vary.

The function f in (2) can be either linear or nonlinear. In this research the more general case of a nonlinear function is considered, but for the sake of completeness, the linear form is presented as follows:

$$x_t = Ax_{t-1} + Bu_t + c_t$$

The aim of control optimization is to find an optimal strategy, i.e. an optimal set of controls $(u_1^*, u_2^*, ..., u_T^*)$ for which $u^* = arg \min_u J$ is true and the whole model (2) holds.

Another important point is the uncertainty or the stochastic in the model. In this paper two kinds of stochastic are considered, namely additive and multiplicative uncertainties. In the first case the vector of additive noise terms ε_t is present in the system. As usual the additive noise is normally (Gaussian) distributed with the expectation 0 and with the given variance-covariance matrix $\Sigma^{\varepsilon\varepsilon}$. In the second case the parameters are uncertain, i.e. a subset (or the whole set) of the elements of the parameter matrices has true constant values but these true values are unknown to the policy-maker. Only the first two moments (mean and variance) of the parameter estimates are known. Assuming θ is the vector of parameters, θ can contain fix parameters as well as uncertain parameters. If at least one parameter in the model is uncertain, the system/model includes uncertainties and these uncertainties are represented by the vector θ. If uncertainty of any kind occurs in the model, the model is named stochastic, otherwise deterministic.

The general form of the optimum control problem (1) - (2) should be adjusted in order to take care of

---
[2]This form of the dynamic problem holds for discrete time only. In the case of continuous time the difference equations are replaced by differential equations $\dot{x}(t) = f(x, u, t)$ and the summation operator 'Σ' in the objective function is replaced by an integral.

the stochastic character of the problem. The intertemporal objective function is formulated in quadratic tracking form[3], which is quite often used in applications of optimum control theory to econometric models. This kind of problem is defined in the following way:

$$J = E\left[\sum_{t=1}^{T} L_t(x_t, u_t)\right], \qquad (3)$$

with

$$L_t(x_t, u_t) = \frac{1}{2}\begin{pmatrix} x_t - \tilde{x}_t \\ u_t - \tilde{u}_t \end{pmatrix}' W_t \begin{pmatrix} x_t - \tilde{x}_t \\ u_t - \tilde{u}_t \end{pmatrix}. \qquad (4)$$

J is the criterion function to be minimized; E is an expectations operator; L_t is a criterion function for period t. x_t is an n-dimensional vector of state variables that describe the state of the economic system at any point in time t. u_t is an m-dimensional vector of control variables, $\tilde{x}_t \in R^n$ and $\tilde{u}_t \in R^m$ are given 'ideal' (desired, target) levels of the state and control variables respectively. T denotes the terminal time period of the finite planning horizon. W_t is an $((n+m) \times (n+m))$ penalty matrix for period t, specifying the relative weights of the state and control variables in the objective function. In a frequent special case, W_t is a matrix including a discount factor α with $W_t = \alpha^{t-1} W$. W_t (or W) is symmetric.

The dynamic system of nonlinear difference equations may be written as

$$x_t = f(x_{t-1}, x_t, u_t, \theta, z_t) + \varepsilon_t, \quad t = 1, ..., T, \qquad (5)$$

where θ is a p-dimensional vector of parameters whose values are assumed to be constant but unknown to the policy-maker (parameter uncertainty). Rather the decision-maker knows only the first two moments (means and variances) of the parameter estimates. z_t denotes an l-dimensional vector of non-controlled exogenous variables, and ε_t is an n-dimensional vector of additive disturbances (system error). θ and ε_t are assumed to be independent random vectors with expectations $\hat{\theta}$ and O_n respectively and covariance matrices $\Sigma^{\theta\theta}$ and $\Sigma^{\varepsilon\varepsilon}$ respectively. f is a vector-valued function and $f^i(.....)$ is the i-th component of $f(.....)$ where $i = 1, ..., n$.

Thus, this paper deals with the algorithm which is designed to provide approximate solutions to optimum control problems with a quadratic objective function (a loss function to be minimized) and a nonlinear multivariate discrete-time dynamic system under additive and parameter uncertainties.

2.2 Literature survey

The aim of my research work is to develop an advanced algorithm for solving the optimum control problems described in subsection 2.1 on the basis of the OPTCON1 algorithm, i.e. an extended version of OPTCON1 that includes an additional control strategy, namely passive learning. Before I discuss the OPTCON1 and the new OPTCON2 algorithms in detail, it is appropriate to mention when and by whom similar concepts and algorithms have been developed.

Although the concepts for dynamic control optimization in economics came from the area of engineering already in the 1920s, they were not widespread in economics for a long time. The first idea to deal with economic problems in the same way as with technical systems goes back to Tustin (1953). Phillips (1954, 1957), Theil (1957) and Holt (1962) also provided some new ideas in this field. Later authors like Athans and Falb (1966), Aoki (1967), Arrow (1968) and Bryson and Ho (1969) published their works, adding more to the policy aspect. In the early 1970s these authors were followed by Kendrick and Taylor (1970),

---
[3]The tracking form is the form of the function that penalizes deviations of the state and control variables from their given target values.

Livesey (1971), Shupp (1972), Friedman and Howrey (1973) and Chow (1975, 1981), who broadened and deepened the theory by adding the concept of nonlinearity. Authors like Bar-Shalom and Sivan (1969), Tse and Athans (1972), Bar-Shalom and Tse (1976a,b), MacRae (1972, 1975), Kendrick (1973) and Chow (1973) also concentrated on uncertainty in the dynamic control optimization theory. Particularly Aoki (1967), Curry (1969), Pitchford and Turnovsky (1977), Chow (1975), Norman (1976) and Kendrick (1981) focused on open-loop feedback strategy in stochastic dynamic control models. Later Coomes (1987), Neck and Matulka (1992, 1994), Amman (1996), Amman and Kendrick (1999), Neck and Karbuz (2000) and Kendrick and Amman (2006) published interesting works on the subject of dynamic stochastic optimum control problems. Although there is a large amount of theoretical knowledge about stochastic control problems, the application of this knowledge is relatively limited; thus, the existing algorithms solve only specific optimum problems. Some algorithms and concepts have been developed by Kendrick (1981) specially for solving dynamic control problems with quadratic objective functions and linear systems taking the additive and multiplicative uncertainties into account. Two versions of Kendrick's algorithm (which are relevant for this research), namely the algorithms with the open-loop and passive learning strategies, are implemented in the computer languages FORTRAN77 and C to the program DUAL.[4] In order to get the open-loop feedback solution the Kalman filter is used to update the stochastic parameters and their covariance matrix in every time period. On the basis of Kendrick's works another algorithm, PLEM, was developed by Coomes (1987). This algorithm also solves the problems with a quadratic objective function and linear system. In contrast to the algorithm mentioned above, PLEM is better adapted to the large models. Another algorithm, OPTNL, developed by Chow and Butters (1977) and Chow (1981), also plays an important role for optimum control theory and is relevant for this work. This algorithm delivers an optimal solution to the dynamic control problem with a quadratic objective function and nonlinear system. The inclusion of nonlinearity in the model is an advantage of OPTNL, but in contrast to other algorithms it only considers additive error in the system and only uses the 'simple' open-loop strategy. The nonlinearity of the system is dealt in OPTNL by using the additional loop that leads to the iterative manner of finding optimal control values. In order to find the start path for the iterative process the Gauss-Seidel or Newton-Raphson method is usually used.

As mentioned above, the basis for this work is the OPTCON1 algorithm developed by R. Neck und J. Matulka. This algorithm combines the techniques for dealing with nonlinearity from OPTNL and the techniques for dealing with additive and multiplicative uncertainties using the open-loop strategy from DUAL/PLEM. OPTCON1 was implemented in GAUSS and applied to several macroeconomic models that are described in the works of Neck and Karbuz (1997), Weyerstrass (1999), Weyerstrass et al. (2001), Weyerstrass and Neck (2007), Neck et al. (2010), Samimi and Tehranchian (2005), Samimi et al. (2010), Tehranchian et al. (2011) and Neck et al. (2011).

Because it is very important for the understanding of further developments, the OPTCON1 algorithm is described in the next chapter.

---

[4] See Kendrick and Coomes (1984).

3 The OPTCON1 algorithm

The basic OPTCON1 algorithm determines approximate solutions to optimum control problems with a quadratic objective function and a nonlinear multivariate dynamic system under additive and parameter uncertainties. The OPTCON1 algorithm is described in detail in Matulka and Neck (1992).

It is well known in stochastic control theory that a general analytical solution to dynamic stochastic optimization problems cannot be achieved even for very simple control problems. The main reason is the so-called dual effect of control under uncertainty, meaning that controls do not only contribute directly to achieving the stated objective but also affect future uncertainty and hence the possibility of indirectly improving on the system's performance by providing better information about the system (see, for instance, Aoki (1989) and Neck (1984)). Given the intricacies of the interplay between control and information, even for very simple stochastic control problems (for example, a linear scalar system with a time horizon of only two periods), an exact analytical solution is impossible. Therefore only approximations to the true optimum for such problems are feasible, with various schemes existing to deal with the problem of information acquisition.

As mentioned above, the OPTCON1 algorithm determines policies belonging to the class of open-loop controls. It either ignores the stochastics of the system altogether (the deterministic solution, identical to the Chow algorithm) or assumes that the stochastics (expectation and covariance matrices of additive and multiplicative disturbances) are given once and for all at the beginning of the planning horizon (the stochastic open-loop solution).

Another important point is the nonlinearity of the system. In search of the approximate solutions OPTCON1 is run iteratively, starting with a tentative path of state and control variables. The tentative path of the control variables is given for the first iteration. In order to find the corresponding tentative path for the state variables, the nonlinear system is solved numerically using the Newton-Raphson method. Alternatively, the Gauss-Seidel method or perturbation methods (e.g. Chen and Zadrozny (2009)) may be used for this purpose.

After the tentative path is found, the iterative approximation of the optimal solution starts. The solution is sought from one time path to another until the algorithm converges or the maximal number of iterations is reached. During this search the system is linearized around the previous iteration's result used as a tentative path and the problem is solved for the resulting time-varying linearized system. The criterion for convergence demands that the difference between the values of current and previous iterations be smaller than a pre-specified number. The approximately optimal solution of the problem for the linearized system is found under the above-mentioned simplifying assumptions about the information pattern; then this solution is used as the tentative path for the next iteration, starting the procedure all over again.

Every iteration, i.e. every solution of the problem for the linearized system, has the following structure: the objective function is minimized using Bellman's principle of optimality[5] to obtain the parameters of the feedback control rule. This uses known results for the stochastic control of LQG problems (the optimization of linear systems with Gaussian noise under a quadratic objective function). A backward recursion over time starts in order to calculate the controls for the first period. Next, the optimal values of the state and the control variables are calculated by applying forward recursion, i.e. beginning with u_1 and x_1 at period 1 and finishing with u_T and x_T at the terminal period T. When the convergence criterion is fulfilled, the solution of the last iteration is taken as the approximately optimal solution to the problem and the algorithm stops. Finally, the value of the objective function is calculated for this solution.

---

[5]Bellman's principle of optimality: "An optimal policy has the property that whatever the initial state and initial decision are, the remaining decisions must constitute an optimal policy with regard to the state resulting from the first decision." See Bellman (1957).

3 THE OPTCON1 ALGORITHM

Figure 1 summarizes the OPTCON1 algorithm:

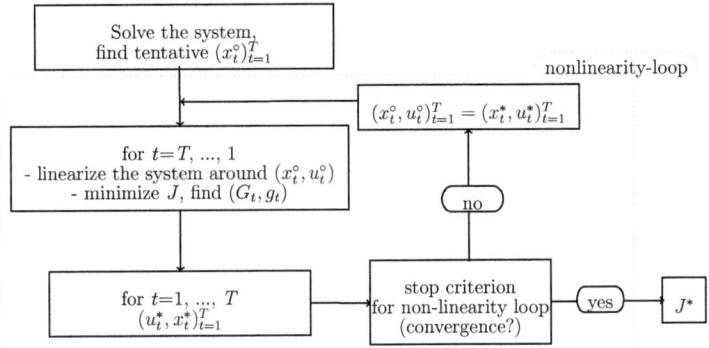

Figure 1: Flow chart of OPTCON1

The next subsection gives a more detailed description of the open-loop solution in the OPTCON1 algorithm.

3.1 Elements of OPTCON1

In order to understand the framework of the OPTCON1 algorithm, which delivers an approximately optimal open-loop solution, some analytical calculations are necessary.

Linearization of the system equations

It is assumed that $\overset{\circ}{x}_{t-1}$, $\overset{\circ}{u}_t$, θ, z_t and ε_t are known. For these values we can compute a value $\overset{\circ}{x}_t$ such that

$$\overset{\circ}{x}_t = f(\overset{\circ}{x}_{t-1}, \overset{\circ}{x}_t, \overset{\circ}{u}_t, \theta, z_t) + \varepsilon_t \tag{6}$$

by the well-known Newton-Raphson (or Gauss-Seidel) approximation algorithm. Then we can linearize the system function $f(...)$ around these reference values and get the following approximative system of equations:

$$x_t = A_t x_{t-1} + B_t u_t + c_t + \xi_t, \quad t=1,...,T \quad, \tag{7}$$

where A_t is an $(n \times n)$-matrix, B_t is an $(n \times m)$-matrix, c_t is an n-dimensional vector and ξ_t is an n-dimensional vector of random error. Following Chow (1975) these matrices and vectors are given as

$$A_t = (I_n - F_{x_t})^{-1} F_{x_{t-1}}, \tag{8}$$

$$B_t = (I_n - F_{x_t})^{-1} F_{u_t}, \tag{9}$$

$$c_t = \overset{\circ}{x}_t - A_t \overset{\circ}{x}_{t-1} - B_t \overset{\circ}{u}_t, \tag{10}$$

$$\xi_t = (I_n - F_{x_t})^{-1} \varepsilon_t, \tag{11}$$

where I_n denotes an $(n \times n)$ identity matrix. For the calculation of the parameters (8)-(11) we require that the first derivatives of the system function with respect to x_{t-1}, x_t, u_t and θ exist and are continuous, and

3 THE OPTCON1 ALGORITHM

we use the following notation:

$$(F_{x_{t-1}})_{i,j} = \frac{\partial f^i(...)}{\partial x_{t-1,j}}, \quad \begin{array}{l} i = 1,...,n \\ j = 1,...,n \end{array} \tag{12}$$

$$(F_{x_t})_{i,j} = \frac{\partial f^i(...)}{\partial x_{t,j}}, \quad \begin{array}{l} i = 1,...,n \\ j = 1,...,n \end{array} \tag{13}$$

$$(F_{u_t})_{i,j} = \frac{\partial f^i(...)}{\partial u_{t,j}}, \quad \begin{array}{l} i = 1,...,n \\ j = 1,...,m \end{array} \tag{14}$$

$$(F_\theta)_{i,j} = \frac{\partial f^i(...)}{\partial \theta_j}, \quad \begin{array}{l} i = 1,...,n \\ j = 1,...,p \end{array} \tag{15}$$

Here $x_{t-1,j}$ denotes the jth element of x_{t-1}. F_{x_t} and $F_{x_{t-1}}$ are $(n \times n)$-matrices, F_{u_t} is an $(n \times m)$-matrix, F_θ is an $(n \times p)$-matrix.

Because of equation (11), the expectation and the covariance matrix of ξ_t, conditional on the information given at $t - 1$, are calculated as

$$E_{t-1}(\xi_t) = (I_n - F_{x_t})^{-1} 0_n = 0_n \tag{16}$$

$$Cov_{t-1}(\xi_t, \xi_t) = (I_n - F_{x_t})^{-1} \Sigma^{\varepsilon\varepsilon}[(I_n - F_{x_t})^{-1}]' \tag{17}$$

Thus, the time-invariant system $f(...)$ is approximated by a time-varying system of linear functions.[6]

Computation of parameter covariances

Because the system includes parameter uncertainty the matrices A_t, B_t and the vector c_t are functions of the random parameter vector θ and, therefore, random themselves. If we write both matrices as collections of their column vectors:

$$A_t = (a_{t,1} \quad ... \quad a_{t,n}), \quad t = 1,...,T$$

$$B_t = (b_{t,1} \quad ... \quad b_{t,m}), \quad t = 1,...,T$$

and notice that these generally nonlinear functions can be approximated by linear functions of θ, then we can write

$$\begin{aligned} a_{t,i} &= D^{a_{t,i}} \theta \\ b_{t,j} &= D^{b_{t,j}} \theta \\ c_t &= D^{c_t} \theta \end{aligned} \tag{18}$$

where $D^{a_{t,i}}$, $D^{b_{t,j}}$ and D^{c_t} are $(n \times p)$-matrices that correspond to column vectors of A_t, B_t and c_t, respectively

$$D^{a_{t,i}} = \begin{bmatrix} \frac{\partial a_{t,1i}}{\partial \theta_1} & \cdots & \frac{\partial a_{t,1i}}{\partial \theta_p} \\ \vdots & \ddots & \vdots \\ \frac{\partial a_{t,ni}}{\partial \theta_1} & \cdots & \frac{\partial a_{t,ni}}{\partial \theta_p} \end{bmatrix}, \quad \begin{array}{l} i = 1,...,n \\ t = 1,...,T \end{array} \tag{19}$$

$$D^{b_{t,j}} = \begin{bmatrix} \frac{\partial b_{t,1j}}{\partial \theta_1} & \cdots & \frac{\partial b_{t,1j}}{\partial \theta_p} \\ \vdots & \ddots & \vdots \\ \frac{\partial b_{t,nj}}{\partial \theta_1} & \cdots & \frac{\partial b_{t,nj}}{\partial \theta_p} \end{bmatrix}, \quad \begin{array}{l} j = 1,...,m \\ t = 1,...,T \end{array} \tag{20}$$

---

[6]Notice that all matrices and vectors defined above depend on the path along which they have been evaluated. Thus, if this path changes the matrices will also change.

3 THE OPTCON1 ALGORITHM

$$D^{c_t} = \begin{bmatrix} \frac{\partial c_{t,1}}{\partial \theta_1} & \cdots & \frac{\partial c_{t,1}}{\partial \theta_p} \\ \vdots & \ddots & \vdots \\ \frac{\partial c_{t,n}}{\partial \theta_1} & \cdots & \frac{\partial c_{t,n}}{\partial \theta_p} \end{bmatrix}, \qquad t = 1, ..., T \tag{21}$$

Notice that the following holds

$$D^{A_t} = \frac{\partial A_t}{\partial \theta} \equiv [vec((D^{a_{t,1}})'), ..., vec((D^{a_{t,j}})'), ..., vec((D^{a_{t,n}})')]$$

$$D^{B_t} = \frac{\partial B_t}{\partial \theta} \equiv [vec((D^{b_{t,1}})'), ..., vec((D^{b_{t,j}})'), ..., vec((D^{b_{t,m}})')]$$

$$D^{c_t} = \frac{\partial c_t}{\partial \theta} \equiv [vec((D^{c_t})')].$$

Moreover, we need the second derivatives in order to compute the matrices $D^{a_{t,j}}$, $D^{b_{t,j}}$ and D^{c_t}. So we define $(np \times n)$-matrices $F_{x_{t-1},\theta}$ and $F_{x_t,\theta}$ as well as an $(np \times m)$-matrix $F_{u_t,\theta}$:

$$F_{x_{t-1},\theta} = \begin{bmatrix} \begin{pmatrix} \frac{\partial f^1}{\partial x_{t-1,1}\partial \theta_1} \\ \vdots \\ \frac{\partial f^1}{\partial x_{t-1,1}\partial \theta_p} \end{pmatrix} & \cdots & \begin{pmatrix} \frac{\partial f^1}{\partial x_{t-1,n}\partial \theta_1} \\ \vdots \\ \frac{\partial f^1}{\partial x_{t-1,n}\partial \theta_p} \end{pmatrix} \\ \vdots & \ddots & \vdots \\ \begin{pmatrix} \frac{\partial f^n}{\partial x_{t-1,1}\partial \theta_1} \\ \vdots \\ \frac{\partial f^n}{\partial x_{t-1,1}\partial \theta_p} \end{pmatrix} & \cdots & \begin{pmatrix} \frac{\partial f^n}{\partial x_{t-1,n}\partial \theta_1} \\ \vdots \\ \frac{\partial f^n}{\partial x_{t-1,n}\partial \theta_p} \end{pmatrix} \end{bmatrix}$$

$$F_{u_t,\theta} = \begin{bmatrix} \begin{pmatrix} \frac{\partial f^1}{\partial u_{t,1}\partial \theta_1} \\ \vdots \\ \frac{\partial f^1}{\partial u_{t,1}\partial \theta_p} \end{pmatrix} & \cdots & \begin{pmatrix} \frac{\partial f^1}{\partial u_{t,m}\partial \theta_1} \\ \vdots \\ \frac{\partial f^1}{\partial u_{t,m}\partial \theta_p} \end{pmatrix} \\ \vdots & \ddots & \vdots \\ \begin{pmatrix} \frac{\partial f^n}{\partial u_{t,1}\partial \theta_1} \\ \vdots \\ \frac{\partial f^n}{\partial u_{t,1}\partial \theta_p} \end{pmatrix} & \cdots & \begin{pmatrix} \frac{\partial f^n}{\partial u_{t,m}\partial \theta_1} \\ \vdots \\ \frac{\partial f^n}{\partial u_{t,m}\partial \theta_p} \end{pmatrix} \end{bmatrix}$$

$F_{x_t,\theta}$ is defined analog to $F_{x_{t-1},\theta}$ with the difference that all occurrences of x_{t-1} are replaced by x_t. Using these derivations we can show the following calculations:[7]

$$D^{A_t} = [(I_n - F_{x_t})^{-1} \otimes I_p][F_{x_t,\theta}A_t + F_{x_{t-1},\theta}], \tag{22}$$

$$D^{B_t} = [(I_n - F_{x_t})^{-1} \otimes I_p][F_{x_t,\theta}B_t + F_{u_t,\theta}], \tag{23}$$

$$d^{c_t} = vec[((I_n - F_{x_t})^{-1}F_\theta)'] - D^{A_t}\overset{\circ}{x}_{t-1} - D^{B_t}\overset{\circ}{u}_t \tag{24}$$

---
[7]The proof for these results is in Appendix 7.1 (Theorem 1).

3 THE OPTCON1 ALGORITHM

We need these forms of matrices for the approximate computation of the covariances and expectations of the parameters of the linearized system. Because of (18) we can write

$$cov_{t-1}(a_{t,i},a_{t,k}) = D^{a_{t,i}}cov_{t-1}(\theta\theta)[D^{a_{t,k}}]' \tag{25}$$

$$cov_{t-1}(a_{t,i},b_{t,j}) = D^{a_{t,i}}cov_{t-1}(\theta\theta)[D^{b_{t,j}}]' \tag{26}$$

$$cov_{t-1}(a_{t,i},c_t) = D^{a_{t,i}}cov_{t-1}(\theta\theta)[D^{c_t}]' \tag{27}$$

$$cov_{t-1}(b_{t,q},b_{t,j}) = D^{b_{t,q}}cov_{t-1}(\theta\theta)[D^{b_{t,j}}]' \tag{28}$$

$$cov_{t-1}(b_{t,j},c_t) = D^{b_{t,j}}cov_{t-1}(\theta\theta)[D^{c_t}]' \tag{29}$$

$$cov_{t-1}(c_t,c_t) = D^{c_t}cov_{t-1}(\theta\theta)[D^{c_t}]' \tag{30}$$

for all $i,k = 1,...,n$; $j,q = 1,...,m$.

Evaluation of some expected values

For the derivation of the OPTCON1 algorithm it becomes necessary to evaluate expectations such as $E_{t-1}(A_t'K_tB_t)$, where K_t denotes a deterministic symmetric $(n \times n)$-matrix. Using the rule $E(XY) = E(X)E(Y) + cov(XY)$ we evaluate $E_{t-1}(A_t'K_tB_t)$ as

$$E_{t-1}(A_t'K_tB_t) = E_{t-1}(A_t)K_tE_{t-1}(B_t) + \Upsilon_t^{AKB}, \tag{31}$$

where $\Upsilon_t^{AKB} = \begin{pmatrix} tr[K_tcov_{t-1}(b_{t,1},a_{t,1})] & \cdots & tr[K_tcov_{t-1}(b_{t,m},a_{t,1})] \\ \vdots & \ddots & \vdots \\ tr[K_tcov_{t-1}(b_{t,1},a_{t,n})] & \cdots & tr[K_tcov_{t-1}(b_{t,m},a_{t,n})] \end{pmatrix}$ is an $(n \times m)$-matrix.

If we define the matrices Υ_t^{AKA}, Υ_t^{BKA}, Υ_t^{BKB}, υ_t^{AKc}, υ_t^{BKc} and υ_t^{cKc} by their elements as

$$[\Upsilon_t^{AKA}]_{i,j} = tr[K_tD^{a_{t,j}}\Sigma^{\theta\theta}(D^{a_{t,i}})'], \quad i = 1,...,n; \quad j = 1,...,n \tag{32}$$

$$[\Upsilon_t^{BKA}]_{i,j} = tr[K_tD^{a_{t,j}}\Sigma^{\theta\theta}(D^{b_{t,i}})'], \quad i = 1,...,m; \quad j = 1,...,n \tag{33}$$

$$[\Upsilon_t^{BKB}]_{i,j} = tr[K_tD^{b_{t,j}}\Sigma^{\theta\theta}(D^{b_{t,i}})'], \quad i = 1,...,m; \quad j = 1,...,m \tag{34}$$

$$[\upsilon_t^{AKc}]_i = tr[K_tD^{c_t}\Sigma^{\theta\theta}(D^{a_{t,i}})'], \quad i = 1,...,n \tag{35}$$

$$[\upsilon_t^{BKc}]_j = tr[K_tD^{c_t}\Sigma^{\theta\theta}(D^{b_{t,j}})'], \quad j = 1,...,m \tag{36}$$

$$[\upsilon_t^{cKc}] = tr[K_tD^{c_t}\Sigma^{\theta\theta}(D^{c_t})'] \tag{37}$$

then the expectations $E_{t-1}(A_t'K_tA_t)$, $E_{t-1}(B_t'K_tA_t)$, $E_{t-1}(B_t'K_tB_t)$, $E_{t-1}(A_t'K_tc_t)$, $E_{t-1}(B_t'K_tc_t)$ and $E_{t-1}(c_t'K_tc_t)$ can be evaluated in an analogous way.

From the "quadratic tracking form" of the objective function to the "general quadratic form"

The "quadratic tracking form" (4) of the objective function is very common in economic policy applications of stochastic control theory. It can be interpreted to require deviations of the state variables and the control variables from their 'ideal' levels to be punished. In order to simplify notation and computation

3 THE OPTCON1 ALGORITHM

the following general quadratic form will be used

$$L_t(x_t, u_t) = \frac{1}{2} \begin{pmatrix} x_t \\ u_t \end{pmatrix}' W_t \begin{pmatrix} x_t \\ u_t \end{pmatrix} + \begin{pmatrix} x_t \\ u_t \end{pmatrix}' \begin{pmatrix} w_t^x \\ w_t^u \end{pmatrix} + w_t^c \tag{38}$$

where

$$\begin{pmatrix} w_t^x \\ w_t^u \end{pmatrix} = -W_t \begin{pmatrix} \tilde{x}_t \\ \tilde{u}_t \end{pmatrix} \quad \text{and} \quad w_t^c = \frac{1}{2} \begin{pmatrix} \tilde{x}_t \\ \tilde{u}_t \end{pmatrix}' W_t \begin{pmatrix} \tilde{x}_t \\ \tilde{u}_t \end{pmatrix}$$

The equivalence between the forms (4) and (38) is shown next.

We start with the general quadratic form (38) and want to achieve the quadratic tracking form (4) by some mathematical calculations.

$$L_t(x_t, u_t) = \frac{1}{2} \begin{pmatrix} x_t \\ u_t \end{pmatrix}' W_t \begin{pmatrix} x_t \\ u_t \end{pmatrix} + \begin{pmatrix} x_t \\ u_t \end{pmatrix}' \begin{pmatrix} w_t^x \\ w_t^u \end{pmatrix} + w_t^c$$

$$= \frac{1}{2} \begin{pmatrix} x_t \\ u_t \end{pmatrix}' W_t \begin{pmatrix} x_t \\ u_t \end{pmatrix} + \begin{pmatrix} x_t \\ u_t \end{pmatrix}' (-W_t) \begin{pmatrix} \tilde{x}_t \\ \tilde{u}_t \end{pmatrix}$$
$$+ \frac{1}{2} \begin{pmatrix} \tilde{x}_t \\ \tilde{u}_t \end{pmatrix}' W_t \begin{pmatrix} \tilde{x}_t \\ \tilde{u}_t \end{pmatrix}$$

$$= \frac{1}{2} \begin{pmatrix} x_t \\ u_t \end{pmatrix}' W_t \begin{pmatrix} x_t - \tilde{x}_t \\ u_t - \tilde{u}_t \end{pmatrix} - \frac{1}{2} \begin{pmatrix} x_t \\ u_t \end{pmatrix}' W_t \begin{pmatrix} \tilde{x}_t \\ \tilde{u}_t \end{pmatrix}$$
$$+ \frac{1}{2} \begin{pmatrix} \tilde{x}_t \\ \tilde{u}_t \end{pmatrix}' W_t \begin{pmatrix} \tilde{x}_t \\ \tilde{u}_t \end{pmatrix}$$

$$= \frac{1}{2} \begin{pmatrix} x_t \\ u_t \end{pmatrix}' W_t \begin{pmatrix} x_t - \tilde{x}_t \\ u_t - \tilde{u}_t \end{pmatrix} - \frac{1}{2} \begin{pmatrix} x_t - \tilde{x}_t \\ u_t - \tilde{u}_t \end{pmatrix}' W_t \begin{pmatrix} \tilde{x}_t \\ \tilde{u}_t \end{pmatrix}$$

$$\stackrel{8}{=} \frac{1}{2} \begin{pmatrix} x_t - \tilde{x}_t \\ u_t - \tilde{u}_t \end{pmatrix}' W_t \begin{pmatrix} x_t - \tilde{x}_t \\ u_t - \tilde{u}_t \end{pmatrix}$$

Approximative solution of the stochastic optimum control problem by OPTCON1

In order to solve the dynamic optimization problem Bellman's principle of optimality should be applied. Bellman's principle of optimality suggests that a local solution of the optimum control problem can be

---
[8] W_t is symmetric.

3 THE OPTCON1 ALGORITHM

obtained over a short time interval τ to form the global solution provided that certain continuity conditions on the solution are satisfied. Adapted to our stochastic optimization problem Bellman's principle has the following form

$$J_t^*(x_{t-1}) = \min_{u_t} E_{t-1}(L_t(x_t, u_t) + J_{t+1}^*(x_t)) \tag{39}$$

where $J_t^*(x_{t-1})$ denotes the loss which is expected in period $t-1$ for the remaining periods $t, ..., T$ if the optimal policy is implemented during these periods.
$J_t^*(x_{t-1})$ can be expressed as a quadratic function of x_{t-1}:[9]

$$J_t^*(x_{t-1}) = \frac{1}{2}x'_{t-1}H_t x_{t-1} + x'_{t-1}h_t^x + h_t^c + h_t^s + h_t^p \tag{40}$$

for all periods $t = 1, ..., T+1$,
where

$$\begin{aligned}
H_t &= \Lambda_t^{xx} - \Lambda_t^{xu}(\Lambda_t^{uu})^{-1}\Lambda_t^{ux} \\
h_t^x &= \lambda_t^x - \Lambda_t^{xu}(\Lambda_t^{uu})^{-1}\lambda_t^u \\
h_t^c &= \lambda_t^c - \frac{1}{2}(\lambda_t^u)'(\Lambda_t^{uu})^{-1}\lambda_t^u \\
h_t^s &= \lambda_t^s, \quad h_t^p = \lambda_t^p
\end{aligned} \tag{41}$$

for all $t = 1, ..., T$.
Because $J_{T+1}^*(x_T) = 0$, hence

$$\begin{aligned}
H_{T+1} &= O_{n \times n}, \quad h_{T+1}^x = O_n \\
h_{T+1}^c &= 0, \quad h_{T+1}^s = 0, \quad h_{T+1}^p = 0
\end{aligned} \tag{42}$$

Next, the auxiliary matrices, vectors and scalar variables are needed and defined as follows:

$$K_t = W_t^{xx} + H_{t+1}, \quad k_t^x = w_t^x + h_{t+1}^x \tag{43}$$

$$\begin{aligned}
\Lambda_t^{xx} &= \Upsilon_t^{AKA} + E_{t-1}(A_t)'K_t E_{t-1}(A_t) \\
\Lambda_t^{ux} &= \Upsilon_t^{BKA} + E_{t-1}(B_t)'K_t E_{t-1}(A_t) + W_t^{ux} E_{t-1}(A_t) \\
\Lambda_t^{xu} &= (\Lambda_t^{ux})' \\
\Lambda_t^{uu} &= \Upsilon_t^{BKB} + E_{t-1}(B_t)'K_t E_{t-1}(B_t) + 2E_{t-1}(B_t)'W_t^{xu} + W_t^{uu}
\end{aligned} \tag{44}$$

$$\begin{aligned}
\lambda_t^x &= v_t^{AKc} + E_{t-1}(A_t)'K_t E_{t-1}(c_t) + E_{t-1}(A_t)'k_t^x \\
\lambda_t^u &= v_t^{BKc} + E_{t-1}(B_t)'K_t E_{t-1}(c_t) + E_{t-1}(B_t)'k_t^x + W_t^{ux}E_{t-1}(c_t) + w_t^u \\
\lambda_t^s &= \frac{1}{2}tr[K_t cov_{t-1}(\xi_t, \xi_t)] + h_{t+1}^s \\
\lambda_t^p &= \frac{1}{2}v_t^{cKc} + h_{t+1}^p \\
\lambda_t^c &= \frac{1}{2}E_{t-1}(c_t)'K_t E_{t-1}(c_t) + E_{t-1}(c_t)'k_t^x + w_t^c + h_{t+1}^c
\end{aligned} \tag{45}$$

Then, after some calculations the following is true

$$J_t(x_{t-1}, u_t) = \frac{1}{2}\begin{pmatrix} x_{t-1} \\ u_t \end{pmatrix}' \begin{pmatrix} \Lambda_t^{xx} & \Lambda_t^{xu} \\ \Lambda_t^{ux} & \Lambda_t^{uu} \end{pmatrix} \begin{pmatrix} x_{t-1} \\ u_t \end{pmatrix} + \begin{pmatrix} x_{t-1} \\ u_t \end{pmatrix}' \begin{pmatrix} \lambda_t^x \\ \lambda_t^u \end{pmatrix} \text{ or}$$
$$+ \lambda_t^c + \lambda_t^s + \lambda_t^p$$

---

[9]See Appendix 7.2 (Theorem 2).

$$J_t(x_{t-1},u_t) = \tfrac{1}{2}(x'_{t-1}\Lambda_t^{xx}x_{t-1}+x'_{t-1}\Lambda_t^{xu}u_t+u'_t\Lambda_t^{ux}x_{t-1}+u'_t\Lambda_t^{uu}u_t)$$
$$+x'_{t-1}\lambda_t^x+u'_t\lambda_t^u+\lambda_t^c+\lambda_t^s+\lambda_t^p \tag{46}$$

Next, the function (46) is to be minimized, i.e. the function $J_t(x_{t-1},u_t)$ is to be differentiated with respect to u_t assuming Λ_t^{uu} is symmetric and positive definite; and this differential it to be set equal to zero.

$$\frac{\partial(J_t(x_{t-1},u_t))}{\partial u_t} = \frac{1}{2}(x'_{t-1}\Lambda_t^{xu}+\Lambda_t^{ux}x_{t-1}+2\Lambda_t^{uu}u_t)+\lambda_t^u = 0$$

Thus we can get the feedback rule for the optimal choice of control u_t^*:

$$u_t^* = -\tfrac{1}{2}(\Lambda_t^{uu})^{-1}x'_{t-1}\Lambda_t^{xu}-\tfrac{1}{2}(\Lambda_t^{uu})^{-1}\Lambda_t^{ux}x_{t-1}-(\Lambda_t^{uu})^{-1}\lambda_t^u$$

$$= -(\Lambda_t^{uu})^{-1}\Lambda_t^{ux}x_{t-1}-(\Lambda_t^{uu})^{-1}\lambda_t^u$$

In another formulation

$$u_t^* = G_t x_{t-1}+g_t \tag{47}$$

where $G_t = -(\Lambda_t^{uu})^{-1}\Lambda_t^{ux}$ and $g_t = -(\Lambda_t^{uu})^{-1}\lambda_t^u$.

3.2 Schematic description of OPTCON1

In this subsection the schematic structure of OPTCON1 is presented. It goes in line with the simplified flow chart presented in Figure 1 and is used as a basic structure for implementing the algorithm.

The following values are given as input:

$f(...)$	system function
$x_0 = \overset{\circ}{x}_0$	initial values of state variables
$(\overset{\circ}{u}_t)_{t=1}^T$	tentative path of control variables
$\hat{\theta}_0 = \hat{\theta}$	expected values of system parameters
$\Sigma_0^{\theta\theta} = \Sigma^{\theta\theta}$	covariance matrix of system parameters
$\Sigma^{\varepsilon\varepsilon}$	covariance matrix of system noise
$(z_t)_{t=1}^T$	path of exogenous variables
$(\tilde{u}_t)_{t=1}^T$	target path for control variables
$(\tilde{x}_t)_{t=1}^T$	target path for state variables
W^{xx}, W^{ux}, W^{uu}	weighting matrices of objective function
α	discount rate of objective function

The output items of the algorithm are the optimal values: $(x_t^*)_{t=1}^T$, $(u_t^*)_{t=1}^T$ and J^*.

Step 1: Compute a tentative state path $(\overset{\circ}{x}_t)_{t=1}^T$ by solving the system of equations $f(...)$ using the Newton-Raphson (or Gauss-Seidel) algorithm given the tentative policy path $(\overset{\circ}{u}_t)_{t=1}^T$ and x_0.

Step 2: *Nonlinearity loop:* repeat the following steps a) to e) until convergence or the maximum number of iterations has been reached.

3 THE OPTCON1 ALGORITHM

Step 2a: The initialization for backward recursion has to be carried out:

$$H_{T+1} = O_{n \times n}, \ h_{T+1}^x = O_n,$$
$$h_{T+1}^c = 0, \ h_{T+1}^s = 0, \ h_{T+1}^p = 0.$$

Step 2b: Backward recursion: for each time period t from T to 1 do the steps [i] - [vii]

[i] Compute the expected values of the parameters of the linearized system of equations:

$$A_t = (I_n - F_{x_t})^{-1} F_{x_{t-1}}, \tag{48}$$

$$B_t = (I_n - F_{x_t})^{-1} F_{u_t}, \tag{49}$$

$$c_t = \overset{\circ}{x}_t - A_t \overset{\circ}{x}_{t-1} - B_t \overset{\circ}{u}_t, \tag{50}$$

$$\Sigma_t^{\xi\xi} = Cov_{t-1}(\xi_t, \xi_t) = (I_n - F_{x_t})^{-1} \Sigma^{\varepsilon\varepsilon} [(I_n - F_{x_t})^{-1}]' \tag{51}$$

where all derivatives are evaluated at the reference values $\overset{\circ}{x}_{t-1}, \overset{\circ}{x}_t, \overset{\circ}{u}_t, \hat{\theta}, z_t,$ and ε_t.
Thus, the time-invariant nonlinear system $f(...)$ is approximated by a time-varying linear system of functions.

[ii] Compute the derivatives of the parameters of the linearized system with respect to θ:

$$D^{A_t} = [(I_n - F_{x_t})^{-1} \otimes I_p][F_{x_t,\theta} A_t + F_{x_{t-1},\theta}], \tag{52}$$

$$D^{B_t} = [(I_n - F_{x_t})^{-1} \otimes I_p][F_{x_t,\theta} B_t + F_{u_t,\theta}], \tag{53}$$

$$d^{c_t} = vec[((I_n - F_{x_t})^{-1} F_\theta)'] - D^{A_t} \overset{\circ}{x}_{t-1} - D^{B_t} \overset{\circ}{u}_t \tag{54}$$

where all derivatives are evaluated at the same reference values as above.

[iii] Compute the influence of the stochastic parameters: determine all the matrices the cells of which are defined by:

$$[\Upsilon_t^{AKA}]_{i,j} = tr[K_t D^{a_{t,j}} \Sigma^{\theta\theta} (D^{a_{t,i}})'], \ i = 1,...,n; \ j = 1,...,n, \tag{55}$$

$$[\Upsilon_t^{BKA}]_{i,j} = tr[K_t D^{a_{t,j}} \Sigma^{\theta\theta} (D^{b_{t,i}})'], \ i = 1,...,m; \ j = 1,...,n, \tag{56}$$

$$[\Upsilon_t^{BKB}]_{i,j} = tr[K_t D^{b_{t,j}} \Sigma^{\theta\theta} (D^{b_{t,i}})'], \ i = 1,...,m; \ j = 1,...,m, \tag{57}$$

$$[\upsilon_t^{AKc}]_i = tr[K_t D^{c_t} \Sigma^{\theta\theta} (D^{a_{t,i}})'], \ i = 1,...,n, \tag{58}$$

$$[\upsilon_t^{BKc}]_i = tr[K_t D^{c_t} \Sigma^{\theta\theta} (D^{b_{t,i}})'], \ i = 1,...,m, \tag{59}$$

$$[\upsilon_t^{cKc}] = tr[K_t D^{c_t} \Sigma^{\theta\theta} (D^{c_t})']. \tag{60}$$

[iv] Convert the objective function from the 'quadratic-tracking' to the 'general quadratic' format:[10]

$$W_t^{xx} = \alpha^{t-1} W^{xx}, \tag{61}$$

$$W_t^{ux} = \alpha^{t-1} W^{ux}, \tag{62}$$

$$W_t^{uu} = \alpha^{t-1} W^{uu}, \tag{63}$$

$$w_t^x = -W_t^{xx} \tilde{x}_t - W_t^{xu} \tilde{u}_t, \tag{64}$$

---

[10] Note that $W_t^{xu} = (W_t^{ux})'$.

3 THE OPTCON1 ALGORITHM

$$w_t^u = -W_t^{ux}\tilde{x}_t - W_t^{uu}\tilde{u}_t, \tag{65}$$

$$w_t^c = \frac{1}{2}\tilde{x}_t' W_t^{xx}\tilde{x}_t + \tilde{u}_t' W_t^{ux}\tilde{x}_t + \frac{1}{2}\tilde{u}_t' W_t^{uu}\tilde{u}_t. \tag{66}$$

[v] Compute the parameters of the function of the expected accumulated loss:

$$\begin{aligned} K_t &= W_t^{xx} + H_{t+1}, \\ k_t^x &= w_t^x + h_{t+1}^x. \end{aligned} \tag{67}$$

$$\begin{aligned} \Lambda_t^{xx} &= \Upsilon_t^{AKA} + A_t' K_t A_t, \\ \Lambda_t^{ux} &= \Upsilon_t^{BKA} + B_t' K_t A_t + W_t^{ux} A_t, \\ \Lambda_t^{xu} &= (\Lambda_t^{ux})', \\ \Lambda_t^{uu} &= \Upsilon_t^{BKB} + B_t' K_t B_t + 2 B_t' W_t^{xu} + W_t^{uu}. \end{aligned} \tag{68}$$

$$\begin{aligned} \lambda_t^x &= v_t^{AKc} + A_t' K_t c_t + A_t' k_t^x, \\ \lambda_t^u &= v_t^{BKc} + B_t' K_t c_t + B_t' k_t^x + W_t^{ux} c_t + w_t^u, \\ \lambda_t^s &= \tfrac{1}{2} tr[K_t \Sigma_t^{\xi\xi}] + h_{t+1}^s, \\ \lambda_t^p &= \tfrac{1}{2} v_t^{cKc} + h_{t+1}^p, \\ \lambda_t^c &= \tfrac{1}{2} c_t' K_t c_t + c_t' k_t^x + w_t^c + h_{t+1}^c. \end{aligned} \tag{69}$$

[vi] Compute the parameters of the feedback rule:

$$\begin{aligned} G_t &= -(\Lambda_t^{uu})^{-1} \Lambda_t^{ux}, \\ g_t &= -(\Lambda_t^{uu})^{-1} \lambda_t^u. \end{aligned} \tag{70}$$

[vii] Compute the parameters of the function of the minimal expected accumulated loss:

$$\begin{aligned} H_t &= \Lambda_t^{xx} - \Lambda_t^{xu}(\Lambda_t^{uu})^{-1}\Lambda_t^{ux}, \\ h_t^x &= \lambda_t^x - \Lambda_t^{xu}(\Lambda_t^{uu})^{-1}\lambda_t^u, \\ h_t^c &= \lambda_t^c - \tfrac{1}{2}(\lambda_t^u)'(\Lambda_t^{uu})^{-1}\lambda_t^u, \\ h_t^s &= \lambda_t^s, \\ h_t^p &= \lambda_t^p. \end{aligned} \tag{71}$$

Step 2c: Start the forward loop: repeat the following steps [i] and [ii] for time periods $t = 1, ..., T$:
[i] The expected optimal policy will be computed using the feedback rule:

$$u_t^* = G_t x_{t-1}^* + g_t.$$

[ii] The optimal values of the state variables have to be calculated using an appropriate system solver (like the Newton-Raphson algorithm):

$$x_t^* = f(x_{t-1}^*, x_t^*, u_t^*, \hat{\theta}, z_t).$$

3 THE OPTCON1 ALGORITHM

Step 2d: Set the new tentative paths for the next iteration:

$$\begin{aligned}(\overset{\circ}{x}_t)_{t=1}^T &= (x_t^*)_{t=1}^T \\ (\overset{\circ}{u}_t)_{t=1}^T &= (u_t^*)_{t=1}^T\end{aligned} \qquad (72)$$

Step 2e: Compute the expected welfare loss:

$$J^* = J_1^* = x_0'H_1x_0 + x_0'h_1^x + h_1^c + h_1^p + h_1^s. \qquad (73)$$

Thus, the OPTCON1 algorithm provides approximate solutions to the optimum control problem (3) - (5).

The aim of my research work is to develop a new algorithm or rather a new, more sophisticated version of the OPTCON algorithm that retains the strengths of the basic version and extends these by the following attributes: using a learning procedure and a user-friendly implementation in a modern computer language. A new OPTCON2 algorithm with the characteristics mentioned above has been created in the scope of this book and will discussed in detail in the next section.

4 The OPTCON2 algorithm

The new version of the algorithm, OPTCON2, incorporates both open-loop and open-loop feedback (passive learning) controls. The idea of passive learning corresponds to actual practice in applied econometrics: at the end of each time period the model builder (and hence the control agent) observes what has happened, that is, the current values of state variables, and uses this information to re-estimate the model and hence improve his/her knowledge of the system. It should be mentioned that two kinds of errors, namely additive (random system errors) and multiplicative ('structural' errors in parameters), are taken into account but not possible specification errors, hence it is assumed that no re-specifications of the model are performed.

The main research aim is to obtain evidence as to whether applying passive learning can indeed improve the final solution. The prediction and optimization procedures for open-loop control assume that the model is not affected by random disturbances occurring during the optimization process. But in reality some random errors will disturb the optimization process. OPTCON2 can deal with two kinds of uncertainties, parameter and system errors. The passive learning strategy implies observing current information and using it in order to adjust the optimization procedure. For the purpose of comparing open-loop and open-loop feedback results, because it is not possible to observe current and true values, one has to resort to Monte Carlo simulations. Large numbers of random time paths for the additive and multiplicative errors are generated, representing what new information could look like in reality. In this way 'quasi-real' observations are created and both types of controls, open-loop and passive learning (open-loop feedback), are compared. The procedure is applied as follows.

M (a number, usually between 100 and 1000) different sets of realizations of random noises $(\varepsilon_t^m)_{t=1}^T$ and μ^m, $m = 1,...,M$, are generated. It is assumed that there is an unknown 'real' model with the 'true' constant parameter vector $\hat{\theta}$. But the policy-maker does not know these 'true' parameters $\hat{\theta}$ and works with the 'wrong' parameters θ^m resulting from the estimates using the realization of the random variable μ^m: $\theta^m = \hat{\theta} + \mu^m$. After this, the following procedure is run for every random scenario: a forward loop is started from 1 to T. In each time period S an (approximately optimal) open-loop solution for the subproblem is determined, i.e. the problem for the time periods from S to T. Then the predicted x_S^* and u_S^* for the time period S are fixed. The assumption is that the policy-maker knows the realized values of the state variables x_S^{a*} at the end of period S, which are, however, disturbed by the additive errors: the difference between $x_S^* = f(x_{S-1}^{a*}, x_S^*, u_S^*, \theta^m, z_S)$ and $x_S^{a*} = f(x_{S-1}^{a*}, x_S^{a*}, u_S^*, \hat{\theta}, z_S) + \varepsilon_S^m$ comes from the realization of the random numbers ε_S^m and μ^m. Next, the new information is used by the policy-maker to update and to adjust the parameter estimate θ^m, whereby the idea of the Kalman filter is used. After that, the same procedure is applied for the remaining subproblem from $S+1$ to T, and so on.

In the next subsection a more detailed description of the steps in the OPTCON2 algorithm is given. The theoretical innovation includes the incorporation of passive learning, which influences the whole algorithm. This requires the creation of additional steps and an adjustment of nearly all equations in terms of that.

4.1 Elements of OPTCON2

Most of the elements of the OPTCON1 algorithm as described in subsection 3.1 can also be used in OPTCON2. Thus, the following elements are nearly identical for both algorithms: 'Linearization of the system equations', 'Computation of parameter covariances', 'Evaluation of some expected values', 'Approximative solution' and 'From the quadratic tracking form of the objective function to the general quadratic form'. The only difference between these elements in OPTCON1 and in OPTCON2 refers to the vector of parameters and system error. These five elements in OPTCON1 are calculated using the

4 THE OPTCON2 ALGORITHM

estimated $\hat{\theta}$ and ε. In contrast, the corresponding calculations in OPTCON2 use the values θ^m and ε^m generated by Monte Carlo simulation. Thus, in the linearization step for example the system $f(...)$ is linearized around the tentative path of $\overset{\circ}{x}_{t-1}$, $\overset{\circ}{x}_t$ and $\overset{\circ}{u}_t$, where the state variable is amongst others a function of θ^m. The following approximative linear system of equations is created:

$$x_t = A_t x_{t-1} + B_t u_t + c_t + \xi_t, \quad t = 1,...,T \quad, \tag{74}$$

where A_t is an $(n \times n)$-matrix, B_t is an $(n \times m)$-matrix, c_t is an n-dimensional vector and ξ_t is an n-dimensional vector of random error. These matrices and vectors are given as

$$A_t = (I_n - F_{x_t})^{-1} F_{x_{t-1}}, \tag{75}$$

$$B_t = (I_n - F_{x_t})^{-1} F_{u_t}, \tag{76}$$

$$c_t = \overset{\circ}{x}_t - A_t \overset{\circ}{x}_{t-1} - B_t \overset{\circ}{u}_t, \tag{77}$$

$$\xi_t = (I_n - F_{x_t})^{-1} \varepsilon_t, \tag{78}$$

where all values of the state variable are influenced by θ^m and ε^m.

For the remaining elements ('Computation of parameter covariances', 'Evaluation of some expected values', 'Approximative solution' and 'From the quadratic tracking form of the objective function to the general quadratic form') adoption to the OPTCON2 algorithm is done in an analogous way, namely the generated θ^m and ε^m are taken into account in order to calculate all necessary values.

A new step which is incorporated into OPTCON2, but which is not included in OPTCON1, is the updating procedure. This element represents the passive learning strategy in OPTCON2 and should be described in detail next.

Updating procedure

In the OPTCON2 algorithm (OLF part) the estimated values of the unknown parameters have to be updated at the end of each time period S ($S = 1, ..., T$) when the new information arrives. The update of parameters will be done using the idea of the Kalman filter[11], namely in our case the predicted parameter values are corrected using the actual (or rather the 'quasi-real') observations x^{a*}.

Actually the updating procedure according to the Kalman filter consists of two distinct phases: prediction and correction. The prediction phase uses the estimate from the previous time period to produce an estimate for the current time period. This predicted estimate is also known as the 'a priori' estimate; although it is an estimate for the current time period, it does not include observation information from the current period. In the correction phase, the current 'a priori' prediction is combined with current observation information to refine the estimate. This improved estimate is termed the 'a posteriori' estimate.

This idea is adapted to OPTCON2 in the following way. In the prediction phase the predicted values of the state variables $\hat{x}_{t/t-1}$, the vector of parameters $\theta^m_{t/t-1}$ and the covariances of the parameters have to be calculated. This phase is reduced because the predicted values $\hat{x}_{t/t-1} = x_t^*$ are calculated in the previous steps of the OPTCON2 algorithm. Thus, at the end of each time period t the values of $\hat{x}_{t/t-1} = x_t^*$ and $\theta^m_{t/t-1} = \theta^m_{t-1}$ are known.[12] In the correction phase the predicted values of the state variables, parameter

---

[11]The basis of the proof for the updating of stochastic parameters is taken from Kendrick (1981). For more information on the Kalman filter see also Kalman (1960), or Welch and Bishop (2006) also available online at http://www.cs.unc.edu/~welch/kalman/

[12]For the reason of generalization the index t (instead of S) is used for the current time period.

4 THE OPTCON2 ALGORITHM

estimates and parameter covariances are updated using the new information obtained at the end of the current period. But this phase is reduced to the calculation of the parameter estimates $\theta^m_{t/t-1}$ and the parameter covariances $\Sigma^{\theta\theta}_{t/t}$, because the 'corrected' values of the state variables $x^{a*}_t = f(x^{a*}_{t-1}, x^{a*}_t, u^*_t, \hat{\theta}, z_t) + \varepsilon^m_t$ for $t = S$ are calculated in the previous step of the algorithm. The important point is that the values of the state variables x^{a*}_t or, to be more precise, the difference between $x^*_t = f(x^{a*}_{t-1}, x^*_t, u^*_t, \theta^m, z_t)$ and x^{a*}_t are used at the end of each time period for the purposes of update.

Thus, the following values have to be calculated within the updating procedure:

1. the predicted values of the covariances (in the prediction phase)
2. the corrected values of the parameter estimates (in the correction phase)
3. the corrected values of the covariances (in the correction phase)

Let us start with the step of the prediction phase where the predicted values of covariances have to be computed.

Step 1: Get the predicted values of covariances $\Sigma^{\theta\theta}_{t/t-1}$, $\Sigma^{xx}_{t/t-1}$ and $\Sigma^{\theta x}_{t/t-1}$.

Following Kendrick's approach we start with the general case

$$\Sigma_{t/t-1} = E\{(x_t - \hat{x}_{t/t-1})(x_t - \hat{x}_{t/t-1})' | \; Y\}$$

where Y is the set of actual ('quasi-real') observations.
First we show that

$$\Sigma_{t/t-1} = F_x \Sigma_{t-1/t-1} F'_x + \Sigma^{\varepsilon\varepsilon} + \frac{1}{2} \sum_i \sum_j e^i e^{j\prime} tr(F^i_{xx} \Sigma_{t-1/t-1} F^j_{xx} \Sigma_{t-1/t-1}) \qquad (79)$$

[13]

We consider the system of equations

$$x_t = f_t(x_{t-1}, u_t, \theta, \ldots) + \varepsilon_t$$

and expand it to the second order (i.e. Taylor expansion) around the path $(x^{a*}_{t-1}, u^*_t) = (\hat{x}_{t-1/t-1}, u^*_t)$:

$$\begin{aligned}
x_t \approx\; & f_t(\hat{x}_{t-1/t-1}, u^*_t, \ldots) + [F_x][x_{t-1} - \hat{x}_{t-1/t-1}] + [F_u][u_t - u^*_t] \\
& + \tfrac{1}{2} \sum_i e^i [x_{t-1} - \hat{x}_{t-1/t-1}]' F^i_{xx} [x_{t-1} - \hat{x}_{t-1/t-1}] \\
& + \tfrac{1}{2} \sum_i e^i [u_t - u^*_t]' F^i_{uu} [u_t - u^*_t] \\
& + \sum_i e^i [u_t - u^*_t]' F^i_{ux} [x_{t-1} - \hat{x}_{t-1/t-1}] + \varepsilon_t
\end{aligned} \qquad (80)$$

where F_x and F_u are the first derivatives of f with respect to x and u respectively; $F^i_{uu} = \frac{\partial^2 f^i}{\partial u^2}$, $F^i_{ux} = \frac{\partial^2 f^i}{\partial u \partial x}$ and $F^i_{xx} = \frac{\partial^2 f^i}{\partial x^2}$ are the second derivatives of f.

Set $u_t = u^*_t$ in (80) and get

$$\begin{aligned}
x_t = &\; f_t(\hat{x}_{t-1/t-1}, u^*_t, \ldots) + F_x [x_{t-1} - \hat{x}_{t-1/t-1}] \\
& + \tfrac{1}{2} \sum_i e^i [x_{t-1} - \hat{x}_{t-1/t-1}]' F^i_{xx} [x_{t-1} - \hat{x}_{t-1/t-1}] + \varepsilon_t
\end{aligned} \qquad (81)$$

---
[13] $\Sigma_{t-1/t-1} = \Sigma$ is the known covariance.

4 THE OPTCON2 ALGORITHM

Taking the expected value of (81) with data through period $t-1$ yields:

$$\hat{x}_{t/t-1} \approx f_t(\hat{x}_{t-1/t-1}, u_t^*, \ldots) + \frac{1}{2} E(\sum_i e^i [x_{t-1} - \hat{x}_{t-1/t-1}]' F_{xx}^i [x_{t-1} - \hat{x}_{t-1/t-1}]) \tag{82}$$

Because of the rule $E(x'Ax) = \hat{x}A\hat{x} + tr(A\Sigma)$ it yields

$$\hat{x}_{t/t-1} \approx f_t(\hat{x}_{t-1/t-1}, u_t^*, \ldots) + \frac{1}{2} \sum_i e^i tr[F_{xx}^i \Sigma_{t-1/t-1}] \tag{83}$$

Using (83) and (81) we write out the following

$$x_t - \hat{x}_{t/t-1} = F_x [x_{t-1} - \hat{x}_{t-1/t-1}] + \frac{1}{2} \sum_i e^i [x_{t-1} - \hat{x}_{t-1/t-1}] F_{xx}^i [x_{t-1} - \hat{x}_{t-1/t-1}]$$
$$+ \varepsilon_t - \frac{1}{2} \sum_i e^i tr[F_{xx}^i \Sigma_{t-1/t-1}] \tag{84}$$

Now we can write $\Sigma_{t/t-1}$ in detail

$$\Sigma_{t/t-1} = E\{(x_t - \hat{x}_{t/t-1})(x_t - \hat{x}_{t/t-1})'\}$$

$$= E\{F_x'[x_{t-1} - \hat{x}_{t-1/t-1}][x_{t-1} - \hat{x}_{t-1/t-1}]'F_x\}$$

$$+ \frac{1}{4} E\{[\sum_i e^i [x_{t-1} - \hat{x}_{t-1/t-1}] F_{xx}^i [x_{t-1} - \hat{x}_{t-1/t-1}]]$$

$$\times [\sum_j e^j [x_{t-1} - \hat{x}_{t-1/t-1}] F_{xx}^j [x_{t-1} - \hat{x}_{t-1/t-1}]]'\} \tag{85}$$

$$+ E\{\varepsilon_t \varepsilon_t'\} + \frac{1}{4} E\{[\sum_i e^i tr[F_{xx}^i \Sigma_{t-1/t-1}]][\sum_j e^j tr[F_{xx}^j \Sigma_{t-1/t-1}]]'\}$$

$$- \frac{1}{2} E\{[\sum_i e^i [x_{t-1} - \hat{x}_{t-1/t-1}] F_{xx}^i [x_{t-1} - \hat{x}_{t-1/t-1}]][\sum_i e^i tr[F_{xx}^i \Sigma_{t-1/t-1}]]'\}$$

Further with $E\{[x_{t-1} - \hat{x}_{t-1/t-1}][x_{t-1} - \hat{x}_{t-1/t-1}]'\} = \Sigma_{t-1/t-1}$ and $E\{\varepsilon_t \varepsilon_t'\} = \Sigma^{\varepsilon\varepsilon}$:

$$\Sigma_{t/t-1} = F_x' \Sigma_{t-1/t-1} F_x + \frac{1}{4} \sum_i \sum_j' E\{[[x_{t-1} - \hat{x}_{t-1/t-1}] F_{xx}^i [x_{t-1} - \hat{x}_{t-1/t-1}]]$$

$$\times [[x_{t-1} - \hat{x}_{t-1/t-1}] F_{xx}^j [x_{t-1} - \hat{x}_{t-1/t-1}]]'\} \tag{86}$$

$$+ \Sigma^{\varepsilon\varepsilon} + \frac{1}{4} [\sum_i e^i tr[F_{xx}^i \Sigma_{t-1/t-1}]][\sum_j e^j tr[F_{xx}^j \Sigma_{t-1/t-1}]]'$$

$$- \frac{1}{2} [\sum_j e^j tr[F_{xx}^j \Sigma_{t-1/t-1}]][\sum_i e^i tr[F_{xx}^i \Sigma_{t-1/t-1}]]'$$

Adapting the following rule[14]

$$E[(x'Ax)(x'Bx)] = 2tr[A\Sigma B\Sigma] + tr[A\Sigma]tr[B\Sigma]$$

---
[14] See Kendrick (1981), Appendix F.

4 THE OPTCON2 ALGORITHM

to (86) one obtains:

$$\Sigma_{t/t-1} = F_x' \Sigma_{t-1/t-1} F_x + \frac{1}{2} \Sigma_i \Sigma_j e^i e^{j'} tr[F_{xx}^i \Sigma_{t-1/t-1} F_{xx}^j \Sigma_{t-1/t-1}]$$

$$+ \frac{1}{4} \Sigma_i \Sigma_j e^i e^{j'} tr[F_{xx}^i \Sigma_{t-1/t-1}] tr[F_{xx}^j \Sigma_{t-1/t-1}]$$

$$+ \Sigma^{\varepsilon\varepsilon} + \frac{1}{4}[\Sigma_i e^i tr[F_{xx}^i \Sigma_{t-1/t-1}]][\Sigma_j e^j tr[F_{xx}^j \Sigma_{t-1/t-1}]]' \qquad (87)$$

$$- \frac{1}{2}[\Sigma_j e^j tr[F_{xx}^j \Sigma_{t-1/t-1}]][\Sigma_i e^i tr[F_{xx}^i \Sigma_{t-1/t-1}]]'$$

Thus proposition (79) is true.

Equation (79) can also be rewritten for the augmented system
$$\hat{z}_{t/t-1} = \begin{bmatrix} \hat{x}_{t/t-1} \\ \cdots \\ \theta_{t/t-1}^m \end{bmatrix} :$$

$$\Sigma_{t/t-1} = F_z' \Sigma_{t-1/t-1} F_z + \Sigma^{\varepsilon\varepsilon} + \frac{1}{2} \sum_i \sum_j e^i e^{j'} tr[F_{zz}^i \Sigma_{t-1/t-1} F_{zz}^j \Sigma_{t-1/t-1}] \qquad (88)$$

and in the matrix form:

$$\Sigma_{t/t-1} = \begin{bmatrix} F_x^x & F_\theta^x \\ F_x^\theta & F_\theta^\theta \end{bmatrix}_{t-1} \begin{bmatrix} \Sigma^{xx} & \Sigma^{x\theta} \\ \Sigma^{\theta x} & \Sigma^{\theta\theta} \end{bmatrix}_{t-1/t-1} \begin{bmatrix} F_x^x & F_\theta^x \\ F_x^\theta & F_\theta^\theta \end{bmatrix}_{t-1}' + \begin{bmatrix} \Sigma^{\varepsilon\varepsilon} & 0 \\ 0 & 0 \end{bmatrix}$$

$$+ \frac{1}{2} \Sigma_{i \in I} \Sigma_{j \in I} e^i e^{j'} tr(\begin{bmatrix} F_{xx}^i & F_{x\theta}^i \\ F_{\theta x}^i & F_{\theta\theta}^i \end{bmatrix} \begin{bmatrix} \Sigma^{xx} & \Sigma^{x\theta} \\ \Sigma^{\theta x} & \Sigma^{\theta\theta} \end{bmatrix}_{t-1/t-1} \qquad (89)$$

$$\times \begin{bmatrix} F_{xx}^j & F_{x\theta}^j \\ F_{\theta x}^j & F_{\theta\theta}^j \end{bmatrix} \begin{bmatrix} \Sigma^{xx} & \Sigma^{x\theta} \\ \Sigma^{\theta x} & \Sigma^{\theta\theta} \end{bmatrix}_{t-1/t-1})$$

Note that
$$f^x = f(\hat{x}_{t-1/t-1}, u_t^*, \ldots), \quad f^\theta = \theta_{t-1}$$
and
$$F_x^x = \frac{\partial f^x}{\partial x}, \quad F_\theta^x = \frac{\partial f^x}{\partial \theta}, \quad F_x^\theta = \frac{\partial f^\theta}{\partial x} = 0, \quad F_\theta^\theta = \frac{\partial f^\theta}{\partial \theta} = I_p$$

4 THE OPTCON2 ALGORITHM

Then the following holds

$$\Sigma_{t/t-1} = \begin{bmatrix} \Sigma^{xx} & \vdots & \Sigma^{x\theta} \\ \dots & & \dots \\ \Sigma^{\theta x} & \vdots & \Sigma^{\theta\theta} \end{bmatrix}_{t/t-1}$$

$$= \begin{bmatrix} F_x^x \Sigma^{xx}(F_x^x)' + F_\theta^x \Sigma^{\theta x}(F_x^x)' + F_x^x \Sigma^{x\theta}(F_\theta^x)' + F_\theta^x \Sigma^{\theta\theta}(F_\theta^x)' & \vdots & F_x^x \Sigma^{x\theta} + F_\theta^x \Sigma^{\theta\theta} \\ \dots\dots\dots\dots\dots\dots\dots\dots\dots\dots\dots\dots\dots\dots\dots\dots\dots\dots & & \dots\dots\dots\dots\dots \\ \Sigma^{\theta x}(F_x^x)' + \Sigma^{\theta\theta}(F_\theta^x)' & \vdots & \Sigma^{\theta\theta} \end{bmatrix}_{t-1/t-1} \quad (90)$$

$$+ \begin{bmatrix} \Sigma^{\varepsilon\varepsilon} & \vdots & 0 \\ \dots & & \dots \\ 0 & \vdots & 0 \end{bmatrix}_{t-1} + \begin{bmatrix} \Pi & \vdots & 0 \\ \dots & & \dots \\ 0 & \vdots & 0 \end{bmatrix}$$

where

$$\Pi = \tfrac{1}{2} \sum_{i \in I} \sum_{j \in I} e^i e^{j'} tr\left(\begin{bmatrix} F_{xx}^i & F_{x\theta}^i \\ F_{\theta x}^i & F_{\theta\theta}^i \end{bmatrix} \begin{bmatrix} \Sigma^{xx} & \Sigma^{x\theta} \\ \Sigma^{\theta x} & \Sigma^{\theta\theta} \end{bmatrix}_{t-1/t-1} \right.$$
$$\left. \times \begin{bmatrix} F_{xx}^j & F_{x\theta}^j \\ F_{\theta x}^j & F_{\theta\theta}^j \end{bmatrix} \begin{bmatrix} \Sigma^{xx} & \Sigma^{x\theta} \\ \Sigma^{\theta x} & \Sigma^{\theta\theta} \end{bmatrix}_{t-1/t-1} \right)$$

$$= \tfrac{1}{2} \sum_{i \in x} \sum_{j \in x} e^i e^{j'} tr\left[\begin{bmatrix} F_{xx}^i \Sigma^{xx} + F_{x\theta}^i \Sigma^{\theta x} & F_{xx}^i \Sigma^{x\theta} + F_{x\theta}^i \Sigma^{\theta\theta} \\ F_{\theta x}^i \Sigma^{xx} + F_{\theta\theta}^i \Sigma^{\theta x} & F_{\theta x}^i \Sigma^{x\theta} + F_{\theta\theta}^i \Sigma^{\theta\theta} \end{bmatrix} \right. \quad (91)$$
$$\left. \times \begin{bmatrix} F_{xx}^j \Sigma^{xx} + F_{x\theta}^j \Sigma^{\theta x} & F_{xx}^j \Sigma^{x\theta} + F_{x\theta}^j \Sigma^{\theta\theta} \\ F_{\theta x}^j \Sigma^{xx} + F_{\theta\theta}^j \Sigma^{\theta x} & F_{\theta x}^j \Sigma^{x\theta} + F_{\theta\theta}^j \Sigma^{\theta\theta} \end{bmatrix} \right]_{t-1/t-1}$$

For the model considered in this paper the following holds:

$$F_{\theta\theta} = 0 \text{ and } \Sigma_{t-1/t-1}^{xx} = \Sigma_{t-1/t-1}^{x\theta} = 0$$

and thus

$$\Pi = \frac{1}{2} \sum_{i \in I} \sum_{j \in I} e^i e^{j'} tr(0) = 0$$

Then the component matrices can be rewritten as

$$\Sigma_{t/t-1}^{xx} = F_{\theta t-1}^x \Sigma_{t-1/t-1}^{\theta\theta} (F_{\theta t-1}^x)' + \Sigma^{\varepsilon\varepsilon} \quad (92)$$

$$\Sigma_{t/t-1}^{x\theta} = (\Sigma_{t-1}^{\theta x})' = F_{\theta t-1}^x \Sigma_{t-1/t-1}^{\theta\theta} \quad (93)$$

$$\Sigma_{t/t-1}^{\theta\theta} = \Sigma_{t-1/t-1}^{\theta\theta} \quad (94)$$

After the predicted values (92) - (94) have been calculated, the corrected values have to be found in the next step.

Step 2: Get the corrected values $\Sigma_{t/t}^{xx}$, $\Sigma_{t/t}^{x\theta}$ and $\Sigma_{t/t}^{\theta\theta}$.
In order to calculate these three matrices the Bayesian method can be used (in the case of normal distribution the Bayesian method estimate is equivalent to the weighted-least-square and maximum-likelihood

4 THE OPTCON2 ALGORITHM

estimate).[15]

We consider the system of 'real' observations

$$y_t = h_t(x_t) + \zeta_t \tag{95}$$

with $E(\zeta) = 0$ and $cov(\zeta) = \Sigma^{\zeta\zeta}$.

We first consider the general case with the measurement error and will later take into account that in our case y_t is given by the actual observation $x_t^{a*} = f(\hat{x}_{t-1/t-1}, u_t^*, \hat{\theta}, ...) + \varepsilon_t$, and thus applied to (95) it yields:

$$h_t(x_t) = x_t \text{ and } \zeta_t = 0$$

We use 'ε_t' here, but in our case it means the generated ε_t^m, where m is a current Monte Carlo run ($m = 1, ..., M$).

Since the conditional distribution $p(x/y)$ is Gaussian, the unconditional maximum-likelihood and weighted-least-square estimates coincide and are given by the conditional mean $E(x|y) = \hat{x}_{t/t}$. Thus according to Kendrick (1981), $p(x/y)$ has the following form

$$p(x/y) = \frac{1}{(2\pi)^{\frac{n}{2}}|\zeta|^{\frac{1}{2}}} \tag{96}$$

$$\times exp\{-\frac{1}{2}([x - \bar{x} - P^{xy}(P^{yy})^{-1}[y - \bar{y}]]\zeta^{-1}[x - \bar{x} - P^{xy}(P^{yy})^{-1}[y - \bar{y}]])\}$$

where $P = \begin{bmatrix} P^{yy} & P^{yx} \\ P^{xy} & P^{xx} \end{bmatrix}$ is the covariance of (y,x), $\bar{x} = E(x)$ and $\bar{y} = E(y)$.

(96) is a normal distribution with mean

$$E(x/y) = \bar{x} + P^{xy}(P^{yy})^{-1}[y - \bar{y}]$$

and covariance

$$\Sigma = P^{xx} - P^{xy}(P^{yy})^{-1}(P^{xy})'$$

After some calculations the following can be obtained:[16]

$$E(x|y) = \bar{x} + [P^{xx}h_x][h_x'P^{xx}h_x + \Sigma^{\zeta\zeta} + \frac{1}{2}\sum_i\sum_j e^i e^{j'} tr[h_{xx}^i P^{xx} h_{xx}^j P^{xx}]]^{-1}[y - \bar{y}] \tag{97}$$

and

$$\Sigma = P^{xx} - [P^{xx}h_x][h_xP^{xx}h_x' + \Sigma^{\zeta\zeta} + \frac{1}{2}\sum_i\sum_j e^i e^{j'} tr[h_{xx}^i P^{xx} h_{xx}^j P^{xx}]]^{-1}[h_xP^{xx}] \tag{98}$$

where

$$\begin{aligned}
P^{xx} &= E\{[x - \bar{x}][x - \bar{x}]'\} \\
P^{xy} &= P^{xx}h_x \\
P^{yy} &= h_xP^{xx}h_x' + \Sigma^{\zeta\zeta} + \frac{1}{2}\sum_i\sum_j e^i e^{j'} tr[h_{xx}^i P^{xx} h_{xx}^j P^{xx}]
\end{aligned} \tag{99}$$

Note h_x and h_{xx} are the first and the second derivatives respectively.

(97) and (98) are the corrections/updates of the mean and covariance in the general case. We just have to adapt equations (97) and (98) to our notation.

With the following notations:

$$P^{xx} = \Sigma_{t/t-1}, \ \Sigma = \Sigma_{t/t}, \ h_x = h_{x,t}$$

---

[15] See Bryson and Ho (1975), Chapter 12.
[16] See Kendrick (1981), Appendix D, equations (D-39) - (D-41).

we obtain from (98) the estimation of Σ:

$$\Sigma_{t/t} = [I - V_t h_{x,t}]\Sigma_{t/t-1} \tag{100}$$

where

$$V_t = \Sigma_{t/t-1} h'_{x,t} \\ \times [h_{x,t}\Sigma_{t/t-1} h'_{x,t} + \Sigma^{\zeta\zeta} + \tfrac{1}{2}\Sigma_i \Sigma_j e^i e^{j'} tr(h^i_{xx}\Sigma_{t/t-1} h^j_{xx}\Sigma_{t/t-1})]^{-1} \tag{101}$$

Moreover we know that in our case

$$h_t(x_t) = I x_t$$

; and ζ_t does not exist. Therefore the observation relationship for the augmented system $\hat{z}_{t/t-1} = \begin{bmatrix} \hat{x}_{t/t-1} \\ \ldots\ldots \\ \theta^m_{t/t-1} \end{bmatrix}$ is:[17]

$$y_t = \begin{bmatrix} I & \vdots & 0 \end{bmatrix} \begin{bmatrix} x_t \\ \theta_t \end{bmatrix}$$

and the derivatives

$$h_z = \begin{bmatrix} h_x & \vdots & h_\theta \end{bmatrix} = \begin{bmatrix} I & \vdots & 0 \end{bmatrix}$$

with $h_x = \dfrac{\partial h}{\partial x}$ and $h_\theta = \dfrac{\partial h}{\partial \theta}$.

The second derivatives are defined as

$$h^i_{zz} = \begin{bmatrix} h^i_{xx} & \vdots & h^i_{x\theta} \\ \ldots & & \ldots \\ h^i_{\theta x} & \vdots & h^i_{\theta\theta} \end{bmatrix} = \begin{bmatrix} 0 & \vdots & 0 \\ \ldots & & \ldots \\ 0 & \vdots & 0 \end{bmatrix}$$

Therefore $h^i_{zz}\Sigma_{t/t-1} = 0$ and thus $tr(h^i_{zz}\Sigma_{t/t-1} h^j_{zz}\Sigma_{t/t-1}) = 0$. Moreover (100) and (101) can be rewritten for the system z_t:

$$\Sigma_{t/t} = [I - V_t h_{z,t}]\Sigma_{t/t-1} \tag{102}$$

where

$$V_t = \Sigma_{t/t-1} h'_{z,t} \\ \times [h_{z,t}\Sigma_{t/t-1} h'_{z,t} + \tfrac{1}{2}\Sigma_i \Sigma_j e^i e^{j'} tr(h^i_{zz}\Sigma_{t/t-1} h^j_{zz}\Sigma_{t/t-1})]^{-1} \tag{103}$$

Using the information about h specified above it yields

$$V_t = \begin{bmatrix} \Sigma^{xx} & \vdots & \Sigma^{x\theta} \\ \ldots & & \ldots \\ \Sigma^{\theta x} & \vdots & \Sigma^{\theta\theta} \end{bmatrix}_{t/t-1} \begin{bmatrix} I \\ \ldots \\ 0 \end{bmatrix} [\begin{bmatrix} I & \vdots & 0 \end{bmatrix} \\ \times \begin{bmatrix} \Sigma^{xx} & \vdots & \Sigma^{x\theta} \\ \ldots & & \ldots \\ \Sigma^{\theta x} & \vdots & \Sigma^{\theta\theta} \end{bmatrix} \begin{bmatrix} I \\ \ldots \\ 0 \end{bmatrix}]^{-1} \tag{104}$$

---

[17]See Kendrick (1981), Appendix L.

4 THE OPTCON2 ALGORITHM

Thus

$$V_t = \begin{bmatrix} \Sigma^{xx}_{t/t-1} \\ \ldots \\ \Sigma^{\theta x}_{t/t-1} \end{bmatrix} [\Sigma^{xx}_{t/t-1}]^{-1} = \begin{bmatrix} I \\ \ldots \\ \Sigma^{\theta x}_{t/t-1}(\Sigma^{xx}_{t/t-1})^{-1} \end{bmatrix} \qquad (105)$$

Substitution of (105) into (102) yields

$$\Sigma_{t/t} = \begin{bmatrix} \Sigma^{xx}_{t/t} & \vdots & \Sigma^{x\theta}_{t/t} \\ \ldots & & \ldots \\ \Sigma^{\theta x}_{t/t} & \vdots & \Sigma^{\theta\theta}_{t/t} \end{bmatrix} = [I - V_t h_{z,t}]\Sigma_{t/t-1}$$

$$= \left(\begin{bmatrix} I & \vdots & 0 \\ \ldots & & \ldots \\ 0 & \vdots & I \end{bmatrix} - \begin{bmatrix} I \\ \ldots \\ \Sigma^{\theta x}_{t/t-1}(\Sigma^{xx}_{t/t-1})^{-1} \end{bmatrix} \begin{bmatrix} I & \vdots & 0 \end{bmatrix} \right) \begin{bmatrix} \Sigma^{xx} & \vdots & \Sigma^{x\theta} \\ \ldots & & \ldots \\ \Sigma^{\theta x} & \vdots & \Sigma^{\theta\theta} \end{bmatrix}_{t/t-1} \qquad (106)$$

$$= \begin{bmatrix} 0 & \vdots & 0 \\ \ldots & & \ldots \\ -\Sigma^{\theta x}_{t/t-1}(\Sigma^{xx}_{t/t-1})^{-1} & \vdots & I \end{bmatrix} \begin{bmatrix} \Sigma^{xx} & \vdots & \Sigma^{x\theta} \\ \ldots & & \ldots \\ \Sigma^{\theta x} & \vdots & \Sigma^{\theta\theta} \end{bmatrix}_{t/t-1}$$

Then the following is true

$$\Sigma^{xx}_{t/t} = 0 \qquad (107)$$

$$\Sigma^{\theta x}_{t/t} = 0 \qquad (108)$$

and

$$\Sigma^{\theta\theta}_{t/t} = \Sigma^{\theta\theta}_{t/t-1} - \Sigma^{\theta x}_{t/t-1}(\Sigma^{xx}_{t/t-1})^{-1}\Sigma^{x\theta}_{t/t-1} \qquad (109)$$

Step 3: Get the corrected $\hat{\theta}_{t/t}$.

Use (97) from the previous step:

$$E(x|y) = \bar{x} + [P^{xx}h_x][h'_x P^{xx} h_x + \Sigma^{\zeta\zeta} + \frac{1}{2}\sum_i\sum_j e^i e^{j'} tr[h^i_{xx} P^{xx} h^j_{xx} P^{xx}]]^{-1}[y - \bar{y}]$$

With the following notation

$$E(x|y) = \hat{x}_{t/t}, \quad \bar{x} = \hat{x}_{t/t-1}, \quad P^{xx} = \Sigma_{t/t-1}$$

we obtain:

$$\hat{x}_{t/t} = \hat{x}_{t/t-1} + V_t[y_t - h_{x,t}\hat{x}_{t/t-1}] \qquad (110)$$

where V_t is defined as in equation (101).

For the augmented system $\hat{z} = \begin{bmatrix} \hat{x} \\ \theta^m \end{bmatrix}$:

$$\hat{z}_{t/t} = \hat{z}_{t/t-1} + V_t[y_t - h_{z,t}\hat{z}_{t/t-1}]$$

where V_t is defined as in equation (105).

4 THE OPTCON2 ALGORITHM

Again with $h_{z,t} = [h_{x,t} \vdots h_{\theta,t}] = [I \vdots 0]$ we get

$$\begin{bmatrix} \hat{x} \\ \theta^m \end{bmatrix}_{t/t} = \begin{bmatrix} \hat{x} \\ \theta^m \end{bmatrix}_{t/t-1} + \begin{bmatrix} I \\ \cdots\cdots\cdots \\ \Sigma^{\theta x}_{t/t-1}(\Sigma^{xx}_{t/t-1})^{-1} \end{bmatrix} [y_t - [I \vdots 0] \begin{bmatrix} \hat{x}_{t/t-1} \\ \theta^m_{t/t-1} \end{bmatrix}]$$

$$= \begin{bmatrix} y_t \\ \cdots\cdots\cdots \\ \theta^m_{t/t-1} + \Sigma^{\theta x}_{t/t-1}(\Sigma^{xx}_{t/t-1})^{-1}[y_t - \hat{x}_{t/t-1}] \end{bmatrix}$$

(111)

Thus, with $y_t = x_t^{a*}$

$$\hat{x}_{t/t} = x_t^{a*}$$

$$\theta^m_{t/t} = \theta^m_{t/t-1} + \Sigma^{\theta x}_{t/t-1}(\Sigma^{xx}_{t/t-1})^{-1}[x_t^{a*} - \hat{x}_{t/t-1}]$$

(112)

Remember that $\hat{x}_{t/t-1}$ is nothing other than x_t^*. The correction of the 'wrong' parameters θ^m is based on the difference between $x_t^* = f(x_{t-1}^{a*}, x_t^*, u_t^*, \theta^m, z_t)$ and $x_t^{a*} = f(x_{t-1}^{a*}, x_t^{a*}, u_t^*, \hat{\theta}, z_t) + \varepsilon_t^m$. It should be mentioned that this difference comes from the realization of the random numbers ε_t^m and μ^m.

So far, all important elements of the new version have been discussed. Next, these elements should be brought together in order to form the new version of the algorithm. In the next section a schematic description of the OPTCON2 algorithm (OLF part) is given.

4.2 Schematic description of OPTCON2

In this subsection the OPTCON2 algorithm (OLF part) is described in a schematic way. On the basis of this rough structure it is implementted in the programming language C#.
The following values are given as input:

$f(...)$	system function
$x_0 = \overset{\circ}{x}_0$	initial values of state variables
$(\overset{\circ}{u}_t)_{t=1}^T$	tentative path of control variables
$\hat{\theta}_0 = \hat{\theta}$	expected values of system parameters
$\Sigma_0^{\theta\theta} = \Sigma^{\theta\theta}$	covariance matrix of system parameters
$\Sigma^{\varepsilon\varepsilon}$	covariance matrix of system noise
$(z_t)_{t=1}^T$	path of exogenous variables
$(\tilde{u}_t)_{t=1}^T$	target path for control variables
$(\tilde{x}_t)_{t=1}^T$	target path for state variables
W^{xx}, W^{ux}, W^{uu}	weighting matrices of objective function
α	discount rate of objective function

The output items of the algorithm are the optimal values: $(x_t^{a*})_{t=1}^T$, $(u_t^*)_{t=1}^T$ and J^*.

STEP I: Compute a tentative state path $(\overset{\circ}{x}_t)_{t=1}^T$ by solving the system of equations $f(.....)$ with the Newton-Raphson algorithm (or Newton-Raphson with line-search expansion), given the initial values x_0 and the tentative policy path $(\overset{\circ}{u}_t)_{t=1}^T$.

STEP II: Generate M paths of random normally distributed system noises $(\varepsilon_t^m)_{t=1}^T$ and parameter noises

4 THE OPTCON2 ALGORITHM

μ^m ($\theta^m = \hat{\theta} + \mu^m$) using the given means and covariance matrices. The given covariance matrices are Cholesky decomposed in order to get the lower-triangular matrices. Applying this to uncorrelated random numbers produces a vector with the covariance properties of the system being modeled.

STEP III: For each independent random scenario with $(\varepsilon_t^m)_{t=1}^T$ and μ^m, i.e. for each Monte Carlo run m ($m = 1, ..., M$), perform the following steps:

Step III-1: For each S from 1 to T, find the open-loop solution for the subproblem $(S, ..., T)$, according to the procedure already implemented in OPTCON1; cf. Section 3 above.

Step III-1a: The *nonlinearity loop* is run until the stop criterion is fulfilled, i.e. until the difference between the values of the current and the previous iteration is smaller than a pre-specified number or the maximal number of iterations has been achieved.

When the stop criterion has been achieved, the approximately optimal solution $(x_t^*, u_t^*)_S^T$ has been found. Then go to the next step **III-1b**. It should be noted that after several runs of the nonlinearity loop, only the solution (x_S^*, u_S^*) for the time period S will be taken as the optimal solution. The calculations of the pairs $(x_{t'}^*, u_{t'}^*)$ for other periods ($t' > S$) have to be done again, taking into account the re-estimated parameters for all periods.

Step III-1b: Calculate the following for one time period S only:

$$x_S^{a*} = f(x_{S-1}^{a*}, x_S^{a*}, u_S^*, \hat{\theta}, z_S) + \varepsilon_S^m.$$

Step III-1c: Update the parameter estimates θ^m using the Kalman filter and the current (realized) values of the variables x_S^{a*}:

[1] Prediction:

$$\hat{x}_{S/S-1} = f(x_{S-1}^{a*}, \hat{x}_{S/S-1}, u_S^*, \theta_{S-1/S-1}^m, z_S) = x_S^*, \quad \theta_{S/S-1}^m = \theta_{S-1/S-1}^m,$$
$$\Sigma_{S/S-1}^{xx} = F_{\theta S-1}^x \Sigma_{S-1/S-1}^{\theta\theta} (F_{\theta S-1}^x)' + \Sigma_S^{\varepsilon\varepsilon}$$
and
$$\Sigma_{S/S-1}^{x\theta} = (\Sigma_{S/S-1}^{\theta x})' = F_{\theta S-1}^x \Sigma_{S-1/S-1}^{\theta\theta}, \quad \Sigma_{S/S-1}^{\theta\theta} = \Sigma_{S-1/S-1}^{\theta\theta}. \tag{113}$$

[2] Correction:

$$\Sigma_{S/S}^{\theta\theta} = \Sigma_{S/S-1}^{\theta\theta} - \Sigma_{S/S-1}^{\theta x} (\Sigma_{S/S-1}^{xx})^{-1} \Sigma_{S/S-1}^{x\theta}$$
and
$$\theta_{S/S}^m = \theta_{S/S-1}^m + \Sigma_{S/S-1}^{\theta x} (\Sigma_{S/S-1}^{xx})^{-1} [x_S^{a*} - x_S^*] \quad \text{and} \quad \hat{x}_{S/S} = x_S^{a*}. \tag{114}$$

Thus the update results in the new values $\theta_{S/S}^m$ and $\Sigma_{S/S}^{\theta\theta}$.

Step III-1d: Set $\theta^m = \theta_{S/S}^m$ and $\Sigma^{\theta\theta} = \Sigma_{S/S}^{\theta\theta}$ and run the procedure for the period $S+1$. This loop is finished when $S = T$ and the approximately optimal open-loop feedback control and state variables for all periods have been found.

Step III-2: Compute the realized (approximately minimal) welfare loss:

$$J^* = \sum_{t=1}^T L_t(x_t^{a*}, u_t^*) \tag{115}$$

29

4 THE OPTCON2 ALGORITHM

with

$$L_t(x_t^{a*}, u_t^*) = \frac{1}{2} \begin{pmatrix} x_t^{a*} - \tilde{x}_t \\ u_t^* - \tilde{u}_t \end{pmatrix}' W_t \begin{pmatrix} x_t^{a*} - \tilde{x}_t \\ u_t^* - \tilde{u}_t \end{pmatrix}. \tag{116}$$

Thus, the optimal open-loop feedback solution $(x_t^{a*}, u_t^*)_{t=1}^T$ is calculated.

The main steps of OPTCON2 are summarized in Figure 2.

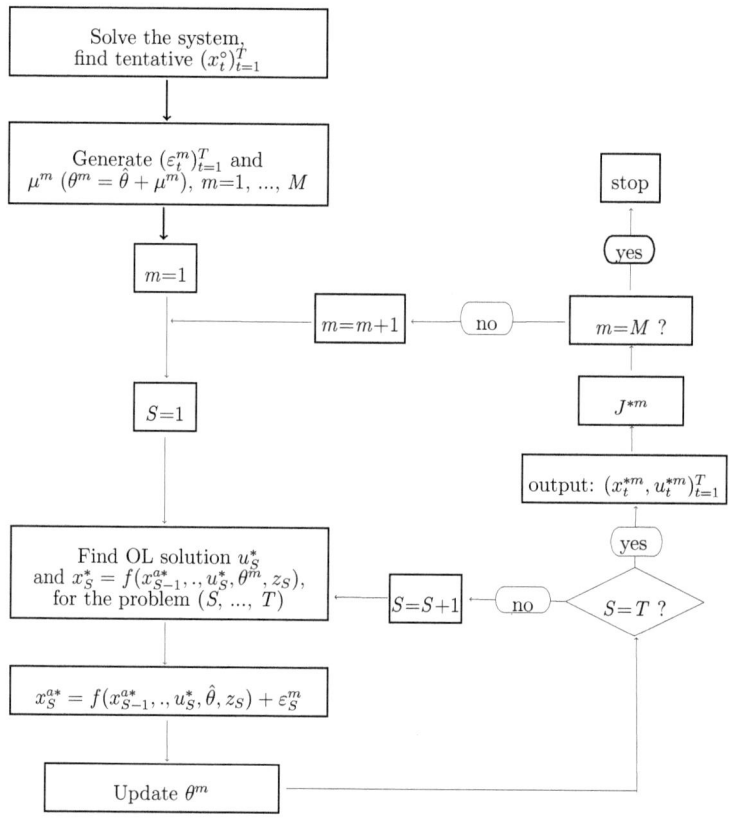

Figure 2: Flow chart of OPTCON2

At this point the theoretical part of my work is complete. The next part deals with the convergence and applicability of the OPTCON2 algorithm.

5 Applications

In this section the practical usefulness of the OPTCON2 algorithm is demonstrated by experiments with four different models, two known from the literature (MacRae and Abel) and two macroeconomic models which were developed for the purposes of this research in collaboration with D. Blueschke and R. Neck. These newly developed models are a nonlinear model of the Slovenian economy (SLOVNL) and a linear macroeconometric model for Slovenia (SLOVL). Comparing the results of the optimum control problems regarding these four models should allow us to draw a conclusion about two control policies, namely open-loop and passive learning.

This section has the following structure. First, in Section 5.1 the macroeconomic model SLOVNL is described and the optimization results of both control strategies are analyzed. Moreover, some additional experiments are run. These experiments include: the comparison of two open-loop solutions, a deterministic one where the variances and covariances of the parameters are ignored, and a stochastic one where the estimated parameter covariance matrix is taken into account; and the inclusion of the weights in the update procedure. The SLOVNL model is based on real data, is larger than the MacRae and Abel models and is nonlinear which makes it more challenging and in some sense more suitable for the aims of this work.

Section 5.2 illustrates the SLOVL model and the results of the OPTCON2 algorithm applied to this linear model. The same experiments as for the SLOVNL model (with the exception of the variation of the weights in the updating procedure) are run. Furthermore, by comparing the results for these two models the impact of nonlinearity on the properties of the optimal solution is analyzed.

In Sections 5.3 and 5.4 we analyze the optimization results of OPTCON2 obtained for two models, namely for the MacRae and Abel models. Both models can also be solved by a comparable algorithm DUAL.[18] DUAL is an algorithm developed by Amman and Kendrick that delivers an optimal solution to linear quadratic deterministic and stochastic control problems. The difference to the OPTCON2 algorithm is that the DUAL algorithm solves optimum control problems for *linear* models only. For this reason DUAL cannot be applied to the nonlinear model SLOVNL. However, by applying both algorithms to the linear econometric model the optimal solutions have to be identical.[19] Thus, the optimization results of the MacRae and Abel models obtained by OPTCON2 can be compared with the solutions of DUAL. The results of these optimum control applications obtained by the DUAL algorithm are assumed to be correct. By comparing the results of the OPTCON2 and the DUAL algorithms it can be checked whether OPTCON2, or more specifically its open-loop feedback strategy, delivers the 'correct' solution.[20]

For all four models the same experiment is run, whereby the properties of the open-loop and the open-loop feedback solutions are compared. There is a problem concerning how to compare both strategies because open-loop controls do not take random disturbances into account during the optimization process. In order to make both strategies comparable, some adjustments to the open-loop controls in the OPTCON2 algorithm are necessary. First, the open-loop controls $(u_t^*)_{t=1}^T$ are calculated for all periods, using the generated θ^m. Looking back to Section 4 the parameters θ^m result from the estimates using the realization of the random variable μ^m and the 'true' constant vector $\hat{\theta}$: $\theta^m = \hat{\theta} + \mu^m$. Then, assuming that the 'true' model is known with the parameters $\hat{\theta}$ and system error ε_t^m, the actual values of the open-loop states $(x_t^{a*} = f(x_{t-1}^{a*}, x_t^{a*}, u_t^*, \hat{\theta}, z_t) + \varepsilon_t^m, t = 1, ..., T)$ are determined, the only information which is available to the decision-maker. So the policies remain unchanged but the state variables are calculated taking the disturbances μ^m and ε^m into account. The open-loop solution adjusted in this way is then

---
[18]See Kendrick and Coomes (1984).
[19]But some slight differences are allowed due to the different computer languages.
[20]The word 'correct' is not exactly the right term for the solution of the OPTCON2 algorithm, because the correct (right) solution is not known due to the properties of the optimum control problem. Nevertheless, I use the word 'correct' in respect to the optimal solution of the OPTCON2 algorithm, in the case when this solution coincides with the DUAL solution.

5 APPLICATIONS

comparable to the open-loop feedback solution. The open-loop feedback solution is determined according to the algorithm sketched in subsection 4.2.
In this way, a comparison of both strategies under simulated uncertainty (disturbances) becomes possible.

5.1 The SLOVNL model

With the aim of illustrating solution concepts and computational techniques, the SLOVNL model is used. This small nonlinear macroeconometric model of the Slovenian economy, called SLOVNL (**SLOV**enian model, **N**on-**L**inear version), consists of 8 equations, 4 of them behavioural and 4 identities. The model includes 8 state variables, 3 control variables, 4 exogenous non-controlled variables and 16 unknown (estimated) parameters. The quarterly data for the time periods 1995:1 to 2006:4 yield 48 observations and admit a full-information maximum likelihood (FIML) estimation of the expected values and the covariance matrices for the parameters and system errors. The start period for the optimization is 2004:1 and the end period is 2006:4 (12 periods).

Model variables used in SLOVNL

Endogenous (state) variables:

$x[1]$:	CR	real private consumption
$x[2]$:	INVR	real investment
$x[3]$:	IMPR	real imports of goods and services
$x[4]$:	STIRLN	short term interest rate
$x[5]$:	GDPR	real gross domestic product
$x[6]$:	VR	real total aggregate demand
$x[7]$:	PV	general price level
$x[8]$:	Pi4	rate of inflation

Control variables:

u[1]	TaxRate	net tax rate
u[2]	GR	real public consumption
u[3]	M3N	money stock, nominal

Exogenous non-controlled variables:

z[1]	EXR	real exports of goods and services
z[2]	IMPDEF	import price level
z[3]	GDPDEF	domestic price level
z[4]	SITEUR	nominal exchange rate SIT/EUR

Model equations:

The first four equations are estimated by FIML, the remaining equations are identities. Standard deviations are given in brackets.

$$CR_t = \underset{(189.7449)}{240.9398} + \underset{(0.1115)}{0.740333} \; CR_{t-1} + \underset{(0.0330)}{0.111727} \; GDPR_t \; (1 - \tfrac{TaxRate_t}{100})$$
$$- \underset{(2.5848)}{1.007353} \; (STIRLN_t - Pi4_t) - \underset{(2.4966)}{4.773533} \; Pi4_t$$

$$INVR_t = \underset{(176.8549)}{75.41731} + \underset{(0.1423)}{0.932211} \; INVR_{t-1} + \underset{(0.0924)}{0.264523} \; (VR_t - VR_{t-1})$$
$$- \underset{(6.9044)}{0.455511} \; (STIRLN_t - Pi4_t) - \underset{(3.1277)}{2.981241} \; Pi4_t$$

5 APPLICATIONS

$$IMPR_t = IMPR_{t-1} + \underset{(0.0724)}{0.826449} (VR_t - VR_{t-1}) - \underset{(18.9336)}{38.14954} SITEUR_t$$

$$STIRLN_t = \underset{(0.1375)}{0.811606} STIRLN_{t-1} - \underset{(0.0008)}{0.000877} \frac{(M3N)_t}{PV_t} \cdot 100$$
$$+ \underset{(0.0026)}{0.002746} GDPR_t$$

$$GDPR_t = CR_t + INVR_t + GR_t + EXR_t - IMPR_t$$

$$VR_t = GDPR_t + IMPR_t$$

$$PV_t = \frac{GDPR_t}{VR_t} \cdot GDPDEF_t + \frac{IMPR_t}{VR_t} \cdot IMPDEF_t$$

$$Pi4_t = \frac{PV_t - PV_{t-4}}{PV_{t-4}} \cdot 100$$

The objective function penalizes deviations of objective variables from their 'ideal' (target) values. The 'ideal' values of the state and control variables ($\tilde{x}_t$ and $\tilde{u}_t$ respectively) are chosen as follows:

Table 1: 'Ideal' values

CR	INVR	IMPR	STIRLN	GDPR	VR	PV	Pi4	TaxRate	GR	M3N
1%	1%	2%	-0.25	1%	1.5%	0.75%	3	25.2	1%	1.75%

The 'ideal' values for most variables are defined in terms of growth rates (denoted by % in Table 1) starting from the last given observation (2003:4). For $Pi4$ and $TaxRate$, constant 'ideal' values are used; for $STIRLN$, a linear decrease of 0.25 per quarter is assumed to be the goal.
The weights for the variables, i.e. matrix W in the objective function, are first chosen as shown in Table 2a ('raw' weights) to reflect the relative importance of the respective variable in the (hypothetical) policy-maker's objective function. These 'raw' weights have to be scaled or normalized according to the levels of corresponding variables to make the weights comparable. To do so, the 'raw' weights are multiplied by normalization coefficients $NC^i = (ML/MA^i)^2$, where ML is the mean of a reference series and MA^i is the mean of the respective series i. The 'correct' weights obtained in this way are shown in Table 2b. The weight matrix is assumed to be constant over time (no discounting).

Table 2: Weights

2a: 'raw' weights		2b: 'correct' weights	
variable	weight	variable	weight
CR	1	CR	3.457677
INVR	1	INVR	12.16323
IMPR	1	IMPR	1.869532
STIRLN	1	STIRLN	216403.9
GDPR	2	GDPR	2
VR	1	VR	0.333598
PV	1	PV	423.9907
Pi4	0	Pi4	0
TaxRate	2	TaxRate	37770.76
GR	2	GR	63.77052
M3N	2	M3N	0.090549

Next, the OPTCON2 algorithm is applied to this optimization problem in order to determine approximately optimal fiscal and monetary policies for Slovenia under the assumed objective function and the

5 APPLICATIONS

econometric model SLOVNL. Two different experiments are run: in experiment 1, two open-loop solutions are compared, a deterministic one where the variances and covariances of the parameters are ignored, and a stochastic one where the estimated parameter covariance matrix is taken into account. In experiment 2, the properties of the open-loop and the open-loop feedback solutions are compared.

Experiment 1: open-loop optimal control

For experiment 1, two different open-loop solutions are calculated: deterministic and stochastic. The deterministic scenario assumes that all parameters of the model are known with certainty. In the stochastic case, the covariance matrix of the parameters as estimated by FIML is used, but no updating of information occurs during the optimization process. This exercise is not meant to determine real optimal policies during the period considered (the quality of the econometric model is not sufficient for this); instead, it should deliver some information about the convergence and the applicability of the OPTCON2 algorithm as implemented in C#.

The results for two variables, real public consumption (GR) and real gross domestic product ($GDPR$), are shown in Figures 3 and 4.[21] Each figure shows the optimization results for the deterministic and the stochastic cases and the 'ideal' and simulated (uncontrolled forecast) values of the respective variables.

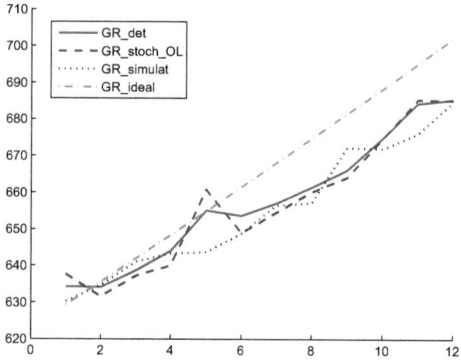

Figure 3: Deterministic vs. stochastic open-loop case (GR)

Figures 3 and 4 show that both the deterministic and the stochastic solutions approximate the 'ideal' values rather well. This is also supported by the values of the objective function, which is 2,618,460.238 in the uncontrolled solution, 904,385.766 in the deterministic solution and 918,296.046 in the stochastic solution, showing a considerable improvement in system performance. An interesting detail is that the deterministic and the stochastic open-loop solutions are very similar, which goes in line with previous findings in a related study by Neck and Karbuz (1995).

In both cases (deterministic and stochastic) the algorithm takes 3 nonlinearity runs to converge. The entire procedure took 2 seconds for the deterministic case and 4 seconds for the stochastic case on a personal computer with a 2 GHz Intel Core 2 Duo CPU and 4GB RAM. The results show that the OPTCON2 algorithm (OL strategy) can be used to determine optimal open-loop controls, at least for small nonlinear econometric models.

---

[21] The results for other objective variables are shown in Appendix 7.4.

5 APPLICATIONS

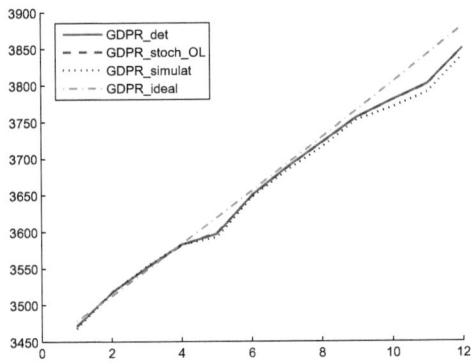

Figure 4: Deterministic vs. stochastic open-loop case (*GDPR*)

Experiment 2: open-loop feedback optimal control

The aim of experiment 2 is to compare open-loop and open-loop feedback controls.

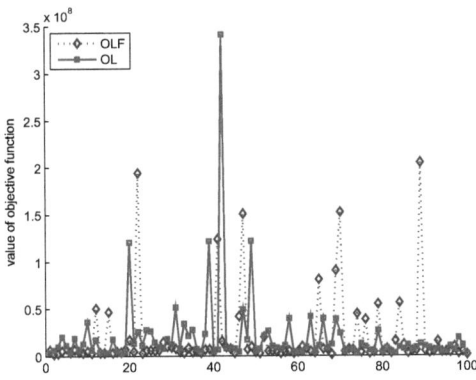

Figure 5: Open-loop vs. open-loop feedback control, value of objective function (100 Monte Carlo runs)

Figure 5 shows the results of a representative Monte Carlo simulation, displaying the value of the objective function (loss) arising from applying OPTCON2 under 100 independent random Monte Carlo runs. Diamonds represent open-loop feedback results and squares open-loop results. One can see from this figure that in most runs the diamonds are below the squares (here: in 66 cases out of 100). This means that open-loop feedback controls give better results (lower values of the cost function) in the majority of the cases investigated. But one can also see that there are many cases where either control scheme results in high losses. After running many simulations having been run (with different numbers of Monte Carlo runs), the findings can be summarized as follows:

- In 60-75 percent of the cases, open-loop feedback controls give better results than open-loop controls.

5 APPLICATIONS

- High losses occur in both the open-loop and the open-loop feedback cases.
- For open-loop controls, high losses seem to be more frequent.

This result is somewhat unexpected because it means that (passive) learning does not necessarily improve the quality of the final results; it may even worsen them. One reason for this is the presence of the two types of stochastic disturbances: additive (random system error) and multiplicative error ('structural' error in the parameters). The decision-maker cannot distinguish between realizations of errors in the parameters and in the equations as he or she just observes the resulting state vector. Based on this information, policy-maker learns about the values of the parameter vector but may be driven away from the 'true' parameter due to the presence of random system error. The possibility of such a diversion can be expected to decrease during the planning horizon as new information (new realizations of the errors) are granted relatively less weight in the updating procedure as time passes by. One possible way out of this dilemma is to introduce weights in the update structure. In particular, in the correction procedure the correction term for the parameter estimate θ is extended by a weighting parameter V_t:

$$\theta_{t/t} = \theta_{t/t-1} + V_t \Sigma^{\theta x}_{t/t-1} (\Sigma^{xx}_{t/t-1})^{-1} [x_t^{a*} - x_t^*], \ 0 < V_t \leq 1. \qquad (117)$$

Adjusting the updating procedure in this way means that better results can be expected under open-loop feedback control. The updating procedure aims at reducing the 'structural' error but can be disturbed by the random system error. Usually, this influence of the random system error can be expected to be especially strong at the beginning of the planning period. Introducing a time-dependent weighting parameter V_t serves to give less weight to revisions called for by learning during the earlier periods of the planning horizon than during the later periods. Different schemes were tried, and the weighted open-loop feedback scheme with the parameter $V_t = \dfrac{t}{T-1}$ gave the best results, so in the simulations presented next this variant was used.

For the purposes of comparing open-loop (OL) and weighted open-loop feedback (wOLF) solutions, a graph, presented in Figure 6, was created showing the values of the objective function against the number of Monte Carlo runs. Diamonds represent wOLF and squares OL results. In this simulation the wOLF policy gives better results: there are more diamonds below squares (77 out of 100). Moreover, only a few wOLF controls result in a very high loss.

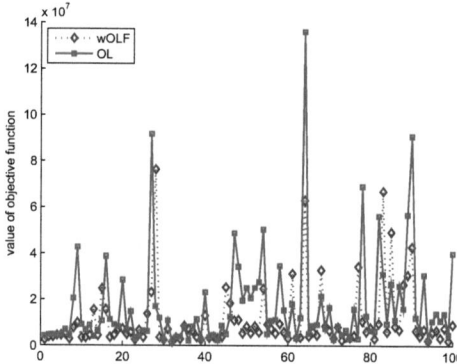

Figure 6: Open-loop vs. weighted open-loop feedback, 100 Monte Carlo runs

5 APPLICATIONS

Another way to show the comparison of open-loop feedback (or weighted open-loop feedback) control with open-loop control was introduced by Amman and Kendrick (1999), viz. scatter diagrams of values of the objective function in different runs. Figures 7 and 8 show such scatter diagrams for 1000 runs in each case. The axes show the values of the objective function under the open-loop strategy (x-axis) and under the (weighted) open-loop feedback strategy (y-axis). One can see that the majority of dots are below the 45 degree line, meaning that OL results in higher losses than OLF (in 66.4% of the runs, Figure 7), and OL results in higher losses than wOLF (in 75.3% of the runs, Figure 8).

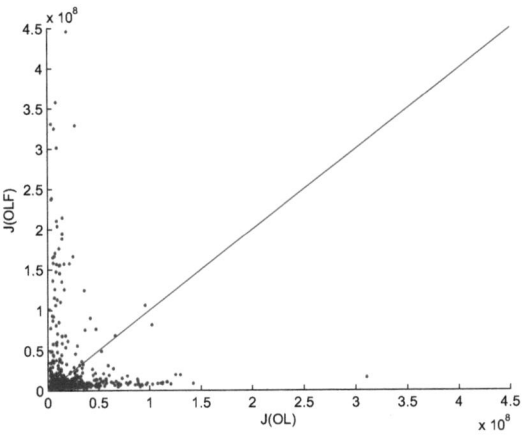

Figure 7: Scatter diagram of OL vs. OLF, 1000 Monte Carlo runs

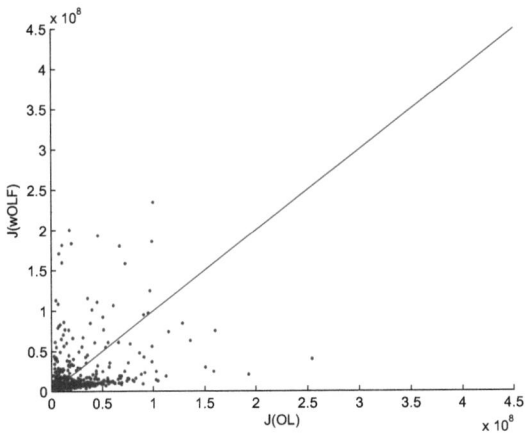

Figure 8: Scatter diagram of OL vs. wOLF, 1000 Monte Carlo runs

After running many simulations the results can be summarized as follows:

5 APPLICATIONS

- In 70-80 percent of the cases considered, weighted open-loop feedback controls give better results than open-loop controls.

- Weighted open-loop feedback controls result in fewer cases of high loss than open-loop controls.

Using a weighting scheme for parameter updating thus increases the number of runs where passive learning controls result in better values of the objective function than the control scheme without learning. Additionally, a decrease in the number of runs with very high losses can be achieved by introducing wOLF controls.

5.2 The SLOVL model

To analyze the impact of the nonlinearity of the system and to test some additional hypotheses, a linear pendant to the SLOVNL model was developed. This 'sister model' is called SLOVL (**SLOV**enian model, Linear version) and consists of 6 equations, 4 of them behavioral and 2 identities. The model includes 6 state variables, 3 control variables, 3 exogenous non-controlled variables and 15 unknown (estimated) parameters. We used the same data base as for SLOVNL and a specification as close as possible to that of SLOVNL in order to make comparisons between the results and convergence properties of the algorithm for a linear and a nonlinear models. Again, we used full-information maximum likelihood (FIML) to estimate the expected values and the covariance matrices for the parameters and the system errors. The starting and the terminal period for the optimization are again 2004:1 and 2006:4 (12 periods).

Model variables used in SLOVL:

Endogenous (state) variables:

$x[1]$	CR	real private consumption
$x[2]$	INVR	real investment
$x[3]$	IMPR	real imports of goods and services
$x[4]$	STIRLN	short term interest rate
$x[5]$	GDPR	real gross domestic product
$x[6]$	VR	real total aggregate demand

Control variables:

u[1]	Taxes	tax revenue
u[2]	GR	real public consumption
u[3]	M3R	money stock, real

Exogenous non-controlled variables:

z[1]	EXR	real exports of goods and services
z[2]	SITEUR	nominal exchange rate SIT/EUR
z[3]	Pi4	rate of inflation

Model equations:

The first four equations are estimated by FIML, the remaining equations are identities. Standard deviations are given in brackets.

5 APPLICATIONS

$$CR_t = 231{,}582776 + 0{,}744522 \; CR_{t-1} + 0{,}111736 \; (GDPR_t - Taxes_t)$$
$$(191{,}99) \quad\quad (0{,}11) \quad\quad\quad\quad (0{,}03)$$
$$- \; 0{,}855137 \; (STIRLN - Pi4) - 4{,}657411 \; Pi4$$
$$(2{,}63) \quad\quad\quad\quad\quad\quad (2{,}50)$$

$$INVR_t = 69{,}965212 + 0{,}936305 \; INVR_{t-1} + 0{,}265119 \; (VR_t - VR_{t-1})$$
$$(176{,}51) \quad\quad (0{,}14) \quad\quad\quad\quad (0{,}09)$$
$$- \; 0{,}292918 \; (STIRLN_t - Pi4_t) - 2{,}869522 \; Pi4_t$$
$$(6{,}90) \quad\quad\quad\quad\quad\quad (3{,}11)$$

$$IMPR_t = IMPR_{t-1} + 0{,}826648 \; (VR_t - VR_{t-1}) - 38{,}158117 \; SITEUR_t$$
$$(0{,}07) \quad\quad\quad\quad\quad\quad (18{,}86)$$

$$STIRLN_t = 0{,}811458 \; STIRLN_{t-1} - 0{,}000877 \; (M3R)_t$$
$$(0{,}14) \quad\quad\quad\quad (0{,}0008)$$
$$+ \; 0{,}002748 \; GDPR_t$$
$$(0.0026)$$

$$GDPR_t = CR_t + INVR_t + GR_t + EXR_t - IMPR_t$$

$$VR_t = GDPR_t + IMPR_t$$

The objective function is analogous to SLOVNL, where the 'ideal' values of the state and control variables ($\tilde{x}_t$ and $\tilde{u}_t$ respectively) are chosen as follows:

Table 3: 'Ideal' values of objective variables, SLOVL

CR	INVR	IMPR	STIRLN	GDPR	VR	Taxes	GR	M3R
1%	1%	2%	-0.25	1%	1.5%	25.2	1%	1.75%

As far as the weights for the variables are concerned, Table 4a shows the 'raw' weights reflecting the relative importance of the respective variable in the objective function, and Table 4b gives the normalized weights.[22] The weight matrix is again assumed to be constant over time.

Table 4: Weights of objective variables, SLOVL

4a: 'raw' weights		4b: 'correct' weights	
variable	weight	variable	weight
CR	1	CR	3.457677
INVR	1	INVR	12.16323
IMPR	1	IMPR	1.869532
STIRLN	1	STIRLN	216403.9
GDPR	2	GDPR	2
VR	1	VR	0.333598
Taxes	2	Taxes	27.52906
GR	2	GR	63.77052
M3R	2	M3R	0.292662

Again two different experiments are run for the SLOVL model as for the SLOVNL model:
1. Deterministic OL and stochastic OL solutions are compared.
2. The properties of the OL and the OLF solutions are compared.

Furthermore, by comparing the results for the SLOVNL and the SLOVL models we want to analyze the impact of nonlinearity on the properties of the optimal solution.

---

[22] The 'raw' weights are normalized in the same way as explained for the SLOVNL model in Subsection 5.1.

5 APPLICATIONS

Experiment 1: open-loop optimal control

Experiment 1, i.e. the calculation of the two different open-loop solutions (a deterministic and a stochastic one), delivers the following results. The results for a control variable, real public consumption (GR), and an endogenous target variable, real gross domestic product ($GDPR$), are shown in Figures 9 and 10. Each figure shows the simulated (uncontrolled forecast) values, the 'ideal' values, and the 'optimal' values (the optimization results for the deterministic and the stochastic cases) of the respective variables.

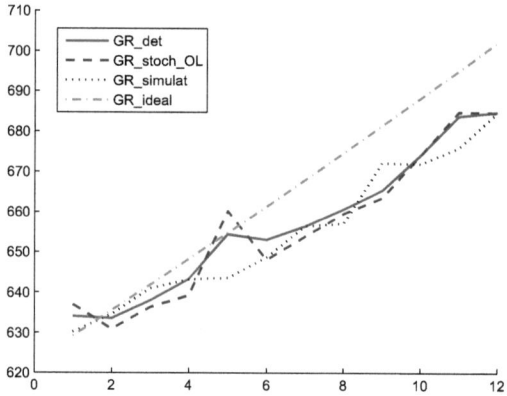

Figure 9: Deterministic and stochastic open-loop policies, GR: SLOVL

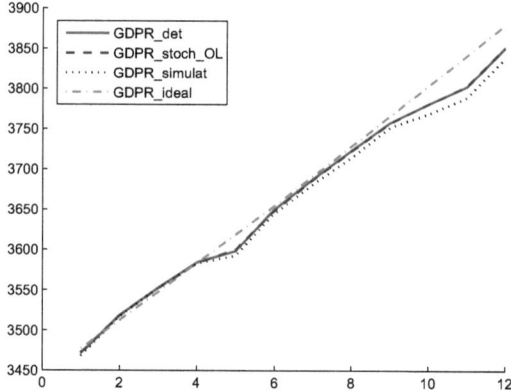

Figure 10: Deterministic and stochastic open-loop policies, $GDPR$: SLOVL

Figures 9 and 10 show that both the deterministic and the stochastic solutions follow the 'ideal' values fairly well but fiscal policies are less expansionary and hence real *GDP* is mostly below its 'ideal' values. In the case of the SLOVL model, the value of the objective function is 3,605,165.99 in the uncontrolled solution, 955,053.796 in the deterministic solution and 968,748.793 in the stochastic solution, showing a significant improvement in system performance. As a reminder, the value of the objective function in the case of the SLOVNL model is 2,618,460.238 in the uncontrolled solution, 904,385.766 in the deterministic solution and 918,296.046 in the stochastic solution, showing a considerable improvement in system performance obtained by optimization and only moderate costs of uncertainty.

An interesting result is that the deterministic and the stochastic open-loop solutions of the SLOVL model are very similar, which goes in line with previous findings in the case of the SLOVNL model.

In order to compare the results of SLOVNL and SLOVL, the graphs for these two models for one example variable (*GR*) are presented in Figure 11. Both solutions are shown together, with the left panel of the figure plotting the solution for SLOVNL and the right panel plotting the solution for SLOVL.[23]

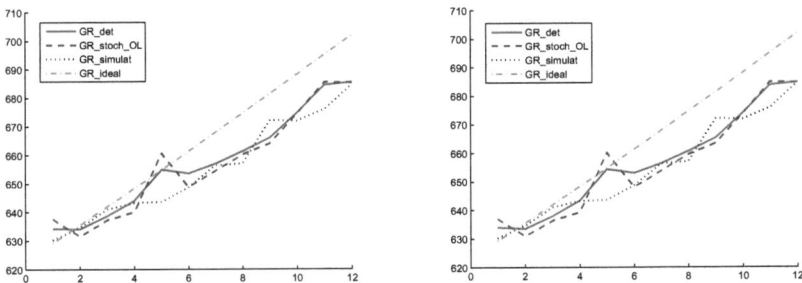

Figure 11: Deterministic and stochastic open-loop policies, *GR*; left: SLOVNL, right: SLOVL

One can see that the SLOVL model is a good 'linear approximation' of the nonlinear SLOVNL model because the results for both models are nearly identical for the variable presented in the graphs above (and for the other variables common to both models). This fact can be used for isolating the impact of nonlinearity in the process of finding the optimum control solution, especially for the case of open-loop feedback policies.

Experiment 2: open-loop feedback optimal control

The aim of experiment 2 consists in comparing open-loop (OL) and open-loop feedback (OLF) optimal stochastic controls. When comparing the two types of (approximately) optimal policies, it must be noted that open-loop controls do not take random disturbances during the optimization process into account. In order to make both policies comparable, some adjustments to the determination of the open-loop controls are necessary. The way of making these adjustments is explained at the beginning of the Section 5. After these adaptations the system is subject to the same stochastic shocks in both cases.

---

[23] The optimal solutions for the remaining variables are given in Appendix 7.4.

5 APPLICATIONS

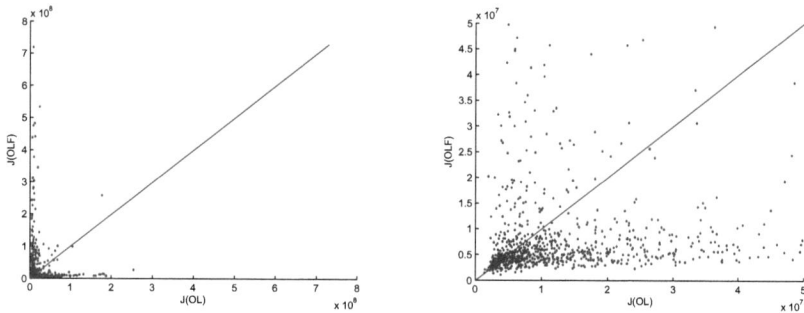

Figure 12: OL and OLF control; SLOVL; left: 'normal', right: 'zoom in'

Figure 12 shows the results of a representative Monte Carlo simulation, displaying the value of the objective function arising from applying OPTCON2 to the SLOVL model, under 1000 independent random Monte Carlo runs. Remember, the objective function is calculated as the sum of the quadratic deviations of the (approximately) optimal values of the objective variables from their 'ideal' values. The graphs plot the values of the objective function for OL policies (x-axis) and OLF policies (y-axis) against each other. In the 'zoom in' panel of the figure, we cut the axes so as to show the mass of the points and omit 'outliers', i.e. results where the value of the objective function becomes extremely large.

One can see that in most cases the values of the objective function for the open-loop feedback solution are smaller than the objective values of the open-loop solution, indicated by a greater mass of dots below the 45 degree line. This means that open-loop feedback controls give better results (lower values of the cost function) in the majority of the cases investigated. For the SLOVL model the OLF policy gives better results than the OL policy, in 65.4% of the cases considered here. Remember for the SLOVNL model it was in 66.4% of the cases.

However, one can also see from these figures (especially in the left-hand panel with a 'normal' view) that there are many cases where either control scheme results in very high losses, indicated by dots which are significantly distant from the main mass of the dots. These cases can be seen even more clearly in Figure 13. This figure shows the same results of the 1000 independent Monte Carlo runs for each model (SLOVNL and SLOVL) separately, but for each Monte Carlo run. The OLF and OL objective function values are plotted in Figure 13 together on the y-axis in each Monte Carlo run, the number of which is shown on the x-axis. Diamonds represent open-loop feedback results and squares represent open-loop results.

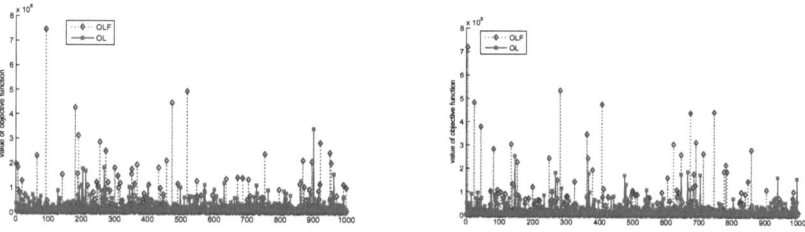

Figure 13: Open-loop vs. open-loop feedback control (left: SLOVNL, right: SLOVL)

Figure 13 illustrates again that most of the results lie within a relatively narrow interval, which means that the optimization procedure works relatively well in most cases. On the other hand, there are also a considerable number of 'outliers' with high losses for both types of policies.

Several additional simulations were run for both models, with different numbers of Monte Carlo runs. Their findings are: in 60-75 percent of the cases open-loop feedback controls give better results than open-loop controls; and high losses ('outliers') occur in both cases (OL and OLF), on average more often in the latter.

Thus, the graphs in this and the previous subsections show that the outliers occur in the linear as well as in the nonlinear model version. But we also obtained some additional information from the Monte Carlo simulations which cannot be seen in the figures. In some of the experiments with 1000 Monte Carlo runs for the SLOVNL model, it turned out that the algorithm did not converge in some runs. In these cases, the algorithm started to diverge, resulting in some non-reasonable or even complex numbers for some variables. In the 1000 Monte Carlo runs experiment applied to the SLOVNL model this happened six times. On the contrary, under the SLOVL model, not one single case of non-convergence occurred out of the 1000 Monte Carlo runs. Thus we arrive at the conclusion that nonlinearity is not the reason for the 'outliers', but it can worsen the problem.

5.3 The MacRae model

The MacRae model, as used by MacRae (1972) and Kendrick (1981), is a purely theoretical model for 2 periods only. The 'one-unknown-parameter' MacRae model includes 1 state variable and 1 control variable. The model does not have exogenous (non-controlled) variables. It consists of 1 equation only:

$$x_t = 3.5 + 0.7\ x_{t-1} - \underset{(0.5)}{0.5}\ u_t + \varepsilon_t \tag{118}$$

One of the parameters is treated as unknown.[24] The objective function penalizes deviations of objective variables from their 'ideal' (desired, target) values. The target values of the state and control variables ($\tilde{x}_t$ and $\tilde{u}_t$ respectively) are assumed to be zero.

The weights for the state and the control variables, i.e. the matrix W in the objective function, are chosen to be 1, which reflects the same importance for all variables. The weight matrix is assumed to be constant over time (no discounting). The optimization horizon consists of 2 periods.

The OPTCON2 algorithm is applied to this optimization problem in order to determine approximately optimal policy under the assumed objective function and the dynamic system given here by equation (118). There are two aims, namely to check:

1. whether the OLF solution of OPTCON2 coincides with the OLF solution of the DUAL algorithm
2. whether the OLF strategy gives better results than the OL strategy

As mentioned before, comparing the two open-loop feedback solutions by the OPTCON2 and DUAL algorithms is done in order to check whether the implementation of OPTCON2 is 'correct' assuming the results of the DUAL algorithm are right. The DUAL algorithm delivers an optimal solution to quadratic linear deterministic and stochastic control problems. Because DUAL can also be used to solve open-loop feedback problems, the results of the OLF strategy obtained for the MacRae model by the OPTCON2 and DUAL algorithms can be compared. To do that a set of random numbers is generated by the DUAL program. Using this set of random numbers the OPTCON2 and the DUAL algorithms each deliver an open-loop feedback solution. Thus, the OLF solutions of both algorithms can be compared. This ex-

---
[24]Standard deviation of the uncertain parameter is given in brackets.

5 APPLICATIONS

periment was conducted several times and the following finding is observed: the optimal values of the state and control variables as well as the values of the objective functions of both algorithms coincide. Let us randomly take a single Monte Carlo run as an example and show the optimal solutions of both algorithms. For the run with $\mu = 0,6051$, $\varepsilon_{t=1} = -0,1963$ and $\varepsilon_{t=2} = -0,2440$ the following values are achieved:

OLF solution (OPTCON2):
$u^*_{t=1} = -0,4560$, $u^*_{t=2} = -0,2035$
$x^*_{t=1} = 3,5317$, $x^*_{t=2} = 5,8300$
$J^* = 23,3553$

OLF solution (DUAL):
$u^*_{t=1} = -0,4560$, $u^*_{t=2} = -0,2036$
$x^*_{t=1} = 3,5317$, $x^*_{t=2} = 5,8300$
$J^* = 23,3555$

We can see that the values of u^*_t and x^*_t for all periods as well as the corresponding function $J^*(u^*_t, x^*_t)$ are nearly identical for both algorithms. The slight difference is acceptable and can be due to the numeric aspects of both programs (OPTCON2 and DUAL) or to the different computer languages.

The fact that two independently developed algorithms deliver the same solution indicates that this solution is 'correct'.

In the next experiment we run 1000 Monte Carlo runs and obtain a set of optimal solutions using the OPTCON2 algorithm. In each run the values of the quadratic objective function by two control strategies are computed. Figure 14 illustrates the optimal results: on the x-axis - the objective values of the open-loop strategy and on the y-axis - the objective values of the open-loop feedback strategy.

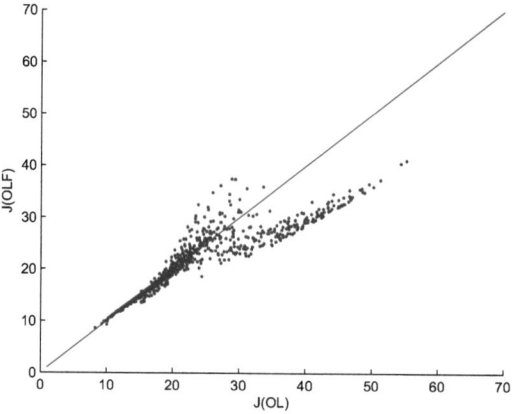

Figure 14: Scatter diagram of OL vs. OLF (OPTCON2)

Looking at Figure 14 one can see that in most runs the dots are below the 45 degree line. This means that the objective function of the open-loop feedback solution has lower values than the objective function of the open-loop control solution in the majority of runs. Numerically we get the following relation: in 68% of the cases OLF controls give better results than OL controls.

Considering the experiments described above one can draw the following conclusion: the implementation and accordingly the performance of the OPTCON2 algorithm (especially the OLF solution) is shown to be 'correct' assuming the performance of the DUAL algorithm is correct. Moreover, the open-loop feedback policy gives better results more often than the open-loop strategy. This implies that passive learning can be advantageous and, in the majority of the cases, the best strategy.

5.4 The Abel model

The next model which is used to demonstrate the applicability and the 'correct' performance of the OPTCON2 algorithm is the Abel model. The model is based on the work of Chow (1967) and Abel (1975). It is a linear macroeconometric model of the U.S. economy that involves two state variables (consumption and investment), two control variables (government obligations and money supply), two exogenous variables (to be put equal to 1) and 10 unknown (estimated) parameters. The optimization time horizon is chosen from 1964:1 to 1965:4. The equations are listed below:[25]

$$CONS_t = \underset{(0.050)}{0.914} CONS_{t-1} - \underset{(0.089)}{0.016} INVEST_{t-1} + \underset{(0.140)}{0.305} GOV_t$$

$$+ \underset{(0.182)}{0.424} MON_t - \underset{(23.345)}{59.437} + \varepsilon_{1t}$$

(119)

$$INVEST_t = \underset{(0.076)}{0.097} CONS_{t-1} + \underset{(0.135)}{0.424} INVEST_{t-1} - \underset{(0.213)}{0.101} GOV_t$$

$$+ \underset{(0.276)}{1.459} MON_t - \underset{(35.440)}{184.766} + \varepsilon_{2t}$$

The matrix W in the objective function includes the values that reflect the relative importance of the respective variable in the (hypothetical) policy-maker's objective function. The weights are chosen as shown in Table 5.

Table 5: Weights

variable	weight
CONS	0.0625
INVEST	1
GOV	1
MON	0.444

The covariance matrix for the parameters and the targets for this model are given in Chow (1967) and Abel (1975).
This model is slightly larger than the MacRae model and the coefficients are estimated from the U.S. data, which makes the experiments more realistic.
Again the OLF solutions of the OPTCON2 and DUAL algorithms are analyzed in order to see whether the passive learning strategy of both algorithms delivers the same results. In this way the correctness of the implementation of OPTCON2 can be checked. The same procedure is conducted as for the MacRae model. A set of random numbers is generated by a Monte Carlo simulation using the DUAL algorithm.

---

[25]Standard deviations of the uncertain parameters are given in brackets.

5 APPLICATIONS

These random numbers are used by the OPTCON2 and the DUAL algorithms in order to find the OLF solutions. Taking a single (randomly chosen) run with a (1×10)- vector μ:
$$\mu = \begin{pmatrix} 0,0133 & 0,0778 & -0,0894 & -0,0839 & 10,8847 & -0,0481 & 0,2342 & 0,0805 & -0,3418 & 37,8425 \end{pmatrix}$$
and additive errors for 7 time periods

$$\varepsilon_1 = \begin{pmatrix} 0,17248 \\ 1,7951 \\ 0,79069 \\ 1,4366 \\ 1,0911 \\ 0,23008 \\ 1,3624 \end{pmatrix}, \quad \varepsilon_2 = \begin{pmatrix} 2,8309 \\ 0,99766 \\ 2,6105 \\ 2,4256 \\ 0,7998 \\ 0,58798 \\ 2,177 \end{pmatrix}$$

the following optimal solutions are calculated:

OLF solution (OPTCON2):

$$u_1^* = \begin{pmatrix} 110,99 \\ 111,89 \\ 112,66 \\ 113,55 \\ 114,40 \\ 115,24 \\ 116,04 \end{pmatrix}, \quad u_2^* = \begin{pmatrix} 143,54 \\ 141,21 \\ 143,12 \\ 142,23 \\ 142,72 \\ 143,84 \\ 145,34 \end{pmatrix}$$

$$x_1^* = \begin{pmatrix} 388,62 \\ 390,12 \\ 391,58 \\ 393,41 \\ 395,20 \\ 396,73 \\ 400,11 \end{pmatrix}, \quad x_2^* = \begin{pmatrix} 90,07 \\ 86,85 \\ 89,95 \\ 89,83 \\ 88,97 \\ 90,11 \\ 94,44 \end{pmatrix}$$

$$J^* = 125,05$$

OLF solution (DUAL):

$$u_1^* = \begin{pmatrix} 111,00 \\ 111,90 \\ 112,69 \\ 113,58 \\ 114,44 \\ 115,28 \\ 116,07 \end{pmatrix}, \quad u_2^* = \begin{pmatrix} 143,54 \\ 141,22 \\ 143,12 \\ 142,24 \\ 142,73 \\ 143,84 \\ 145,31 \end{pmatrix}$$

$$x_1^* = \begin{pmatrix} 388,63 \\ 390,13 \\ 391,59 \\ 393,43 \\ 395,23 \\ 396,76 \\ 400,13 \end{pmatrix}, \quad x_2^* = \begin{pmatrix} 90,07 \\ 86,85 \\ 89,95 \\ 89,84 \\ 88,97 \\ 90,10 \\ 94,38 \end{pmatrix}$$

$$J^* = 124,96$$

Running an experiment with a single (randomly chosen) Monte Carlo run several times results in the conclusion that the open-loop feedback strategy in the OPTCON2 and in the DUAL algorithms delivers

5 APPLICATIONS

nearly identical[26] optimal values of the control and the state variables. This illustrates that the implementation and the performance of the OPTCON2 algorithm seems to be correct. This conclusion coincides with the conclusion from the Subsection 5.3, namely with the results for the MacRae model. This analogy allows us to draw a conclusion that the results in the case of the MacRae model are not a fortunate coincidence but a regularity. Thus, based on these results one can say that the implementation and the performance of the OPTCON2 algorithm are 'correct'.

Next, we run an experiment with 1000 Monte Carlo runs applying the OPTCON2 algorithm to the Abel model and compare two control strategies: open-loop and open-loop feedback. By doing so one can see which control strategy gives better results and what is the ratio of runs with the advantageous policy. Figure 15 illustrates the results of this experiment. The greater number of points lie under the 45 degree line, which indicates that the objective values of the OLF solutions are smaller than the objective values of the OL solutions in the majority of the runs, namely in 74%.

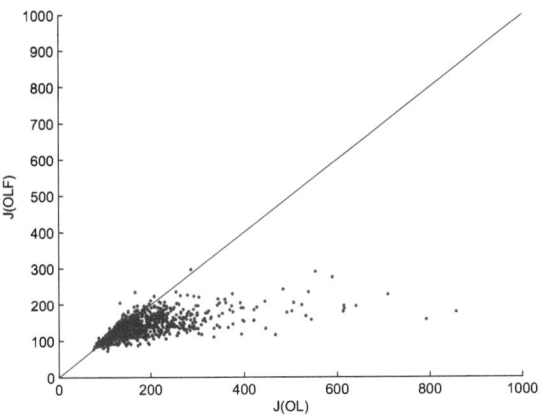

Figure 15: Scatter diagram OL versus OLF, OPTCON2

Thus, the OPTCON2 results of the Abel model and the results of the models described above coincide in the sense that the passive learning policy gives better results than the OL policy in the majority of cases. Altogether it indicates that in many situations the passive learning strategy is an advantageous control strategy that is prudent to use.

---

[26] As said before, the slight difference between the optimal solutions of both algorithms (OPTCON2 and DUAL) is acceptable and can be due to the numeric aspect of both programs or to the different computer languages.

6 Conclusion

The aim of my research was to develop an extension of the OPTCON algorithm, namely a new version of the algorithm that includes two kinds of control strategy, open-loop and passive learning. The theoretical developments, the implementation in computer language *C#* as well as the application of the OPTCON2 algorithm to some macroeconomic models are performed. The theoretical part of the work gives all the details and elements of the algorithm which are necessary in order to understand the innovations. The innovations are made in respect to including an alternative control strategy, open-loop feedback, in the algorithm. Using this strategy it is possible to learn about stochastically disturbed parameters during the control process in each time period using newly arrived information. The learning effect is received by update of uncertain parameters due to the idea of the Kalman filter. All the theoretical calculations and proofs show that the OPTCON2 algorithm delivers a 'correct' numerical solution in terms of mathematics.

The application part of my work demonstrates the usefulness of the new version of the algorithm. The applicability and the convergence of the OPTCON2 algorithm are shown on the four models, namely SLOVNL, SLOVL, MacRae and Abel. Two theoretical models (MacRae and Abel) are also used with Kendrick's algorithm DUAL in order to solve optimum control problems allowing for linear systems. Thus, I could compare the results of the OPTCON2 algorithm and the results of the DUAL algorithm using OLF control schemes in order to check the performance of the OPTCON2 algorithm. Because both algorithms deliver the same results for these models one can say that the implementation of the OPTCON2 algorithm (in the computer language *C#*) and accordingly its performance are 'correct' in regard to the correctness of Kendrick's DUAL algorithm. The second aim of applying the OPTCON2 algorithms to the MacRae and Abel models was the comparison of the open-loop feedback and the open-loop optimal solutions. It is shown that open-loop feedback controls give better results in the majority of the cases investigated for the OPTCON2 algorithm.

But the more important comparison of two control schemes was made on the basis of two macroeconometric models, SLOVNL and SLOVL, which were created for the purpose of this book. The OL (stochastic) solution is compared with the OLF solution in order to examine the effect of parameter updating on the resulting value of the criterion. The results indicate that open-loop feedback control outperforms open-loop control in the majority of the cases investigated for two econometric models of Slovenia. But it suffers from a problem of 'outliers' which is present for both policy schemes and for both models. One of the possibilities to solve or at least to reduce the 'outliers' problem is to include the weights in the update procedure. It is shown that this tactic improves the results, namely the number of outliers in the solutions of the OPTCON2 algorithm (OLF part) is reduced by including the updating weights.

By comparing the optimal solutions for the nonlinear SLOVNL model and for the linear 'sister' SLOVL model the following conclusion is drawn about the impact of nonlinearity on the results . It was demonstrated that the nonlinearity of the model does not impact on the number of outliers, but can worsen the ability of the algorithm to converge.

Altogether, an open-loop feedback strategy can give better solutions than an open-loop strategy. The application of the OPTCON2 algorithm to the four models described in this book reveals that in 60% - 75% of the runs passive learning gives better results.

7 Appendix

7.1 Theorem 1

Proof:
We start with MacRae's definition of the differentiation of matrix Y whose elements are functions of an $(m \times n)$-matrix X:[27]

$$\frac{\partial Y}{\partial X} = Y \otimes \frac{\partial}{\partial X}$$

Also we use the results of MacRae's theorems, namely the product rule:

$$\frac{\partial YZ}{\partial X} = \frac{\partial Y}{\partial X}(Z \otimes I_n) + (Y \otimes I_m)\frac{\partial Z}{\partial X}$$

and the inverse rule:

$$\frac{\partial Y^{-1}}{\partial X} = -(Y^{-1} \otimes I_m)\left(\frac{\partial Y}{\partial X}\right)(Y^{-1} \otimes I_n)$$

where Y (non-singular) and Z are the matrices whose elements are functions of the $(m \times n)$ matrix X; and the product YZ is well-defined.

1) We know that $A_t = (I_n - F_{x_t})^{-1} F_{x_{t-1}}$. Applying the product rule to our case with $Y = I_n - F_{x_t}$, $Z = F_{x_{t-1}}$ and a $(p \times 1)$-vector $X = \theta$ we obtain

$$D^{A_t} = \frac{\partial A_t}{\partial \theta}$$

$$= \frac{\partial [(I_n - F_{x_t})^{-1} F_{x_{t-1}}]}{\partial \theta} \quad (120)$$

$$= \frac{\partial [(I_n - F_{x_t})^{-1}]}{\partial \theta} F_{x_{t-1}} + [(I_n - F_{x_t})^{-1} \otimes I_p]\frac{\partial [F_{x_{t-1}}]}{\partial \theta}$$

As I_n does not depend on θ we get with the inverse rule

$$\frac{\partial [(I_n - F_{x_t})^{-1}]}{\partial \theta} = -[(I_n - F_{x_t})^{-1} \otimes I_p]\frac{\partial (I_n - F_{x_t})}{\partial \theta}((I_n - F_{x_t})^{-1} \otimes I_1) \quad (121)$$

$$= [(I_n - F_{x_t})^{-1} \otimes I_p]\frac{\partial F_{x_t}}{\partial \theta}(I_n - F_{x_t})^{-1}$$

Setting (121) in (120) and using $(I_n - F_{x_t})^{-1} F_{x_{t-1}} = A_t$ we get

$$D^{A_t} = [(I_n - F_{x_t})^{-1} \otimes I_p] F_{x_t,\theta}(I_n - F_{x_t})^{-1} F_{x_{t-1}} + [(I_n - F_{x_t})^{-1} \otimes I_p] F_{x_{t-1},\theta}$$
$$= [(I_n - F_{x_t})^{-1} \otimes I_p][F_{x_t,\theta} A_t + F_{x_{t-1},\theta}]$$

2) We know that $B_t = (I_n - F_{x_t})^{-1} F_{u_t}$.
Again with the product rule we obtain

$$D^{B_t} = \frac{\partial B_t}{\partial \theta}$$

$$= \frac{\partial [(I_n - F_{x_t})^{-1} F_{u_t}]}{\partial \theta} = \frac{\partial [(I_n - F_{x_t})^{-1}]}{\partial \theta} F_{u_t} + [(I_n - F_{x_t})^{-1} \otimes I_p]\frac{\partial [F_{u_t}]}{\partial \theta}$$

Using (121) and $(I_n - F_{x_t})^{-1} F_{u_t} = B_t$ we get

---
[27]See MacRae (1975).

7 APPENDIX

$$D^{B_t} = [(I_n - F_{x_t})^{-1} \otimes I_p] F_{x_t,\theta} (I_n - F_{x_t})^{-1} F_{u_t} + [(I_n - F_{x_t})^{-1} \otimes I_p] F_{u_t,\theta}$$

$$= [(I_n - F_{x_t})^{-1} \otimes I_p][F_{x_t,\theta} B_t + F_{u_t,\theta}]$$

3) We know that $c_t = x_t + A_t x_{t-1} - B_t u_t$. Then

$$D^{c_t} = \frac{\partial c_t}{\partial \theta} = \frac{\partial x_t}{\partial \theta} - \frac{\partial [A_t x_{t-1}]}{\partial \theta} - \frac{\partial [B_t u_t]}{\partial \theta}$$

Applying the product rule twice yields:

$$\frac{\partial c_t}{\partial \theta} = \frac{\partial x_t}{\partial \theta} - D^{A_t} x_{t-1} - (A_t \otimes I_p) \frac{\partial x_{t-1}}{\partial \theta} - D^{B_t} u_t - (B_t \otimes I_p) \frac{\partial u_t}{\partial \theta}$$

Because x_{t-1} and u_t do not depend on θ ($\frac{\partial x_{t-1}}{\partial \theta} = 0$ and $\frac{\partial u_t}{\partial \theta} = 0$), we obtain

$$\frac{\partial c_t}{\partial \theta} = \frac{\partial x_t}{\partial \theta} - D^{A_t} x_{t-1} - D^{B_t} u_t$$

x_t does depend on θ via the system function. Because of that we get:

$$\frac{\partial c_t}{\partial \theta} = vec[((I_n - F_{x_t})^{-1} F_\theta)'] - D^{A_t} x_{t-1} - D^{B_t} u_t$$

where vectorization is necessary to reshape the first derivatives of the system equation (for which we have used the more common notation) to conform to the way derivatives are defined by MacRae. The functions mentioned above have been assumed to be evaluated along the reference path.

Thus, all three statements are proved.

7.2 Theorem 2

To show:
$$J_t^*(x_{t-1}) = \frac{1}{2} x_{t-1}' H_t x_{t-1} + x_{t-1}' h_t^x + h_t^c + h_t^s + h_t^p$$

for all $t = 1, ..., T+1$

Proof:

We do that by complete induction.
Obviously, this is true for $t = T+1$.[28]

We do the induction step ($t \to t-1$):
if $J_{t+1}^*(x_t)$ can be expressed as a quadratic function of x_t, then $J_t^*(x_{t-1})$ will also be quadratic in x_{t-1}.
Thus, we presume that

$$J_{t+1}^*(x_t) = \frac{1}{2} x_t' H_{t+1} x_t + x_t' h_{t+1}^x + h_{t+1}^c + h_{t+1}^s + h_{t+1}^p \qquad (122)$$

holds.
Then, using the general quadratic form of the objective function (38), the assumption (122) and the definition (43) for K_t and k_t^x, we get

---

[28] $J_{T+1}^*(x_T) = 0$, because we do not count any losses after the end of the planning period. $\frac{1}{2} x_T' H_{T+1} x_T + x_T' h_{T+1}^x + h_{T+1}^c + h_{T+1}^s + h_{T+1}^p = 0$, because of (42).

7 APPENDIX

$$L_t(x_t,u_t)+J_{t+1}^*(x_t) = \frac{1}{2}\begin{pmatrix} x_t \\ u_t \end{pmatrix}'\begin{pmatrix} K_t & W_t^{xu} \\ W_t^{ux} & W_t^{uu} \end{pmatrix}\begin{pmatrix} x_t \\ u_t \end{pmatrix} + \begin{pmatrix} x_t \\ u_t \end{pmatrix}'\begin{pmatrix} k_t^x \\ w_t^u \end{pmatrix}$$

$$+w_t^c + h_{t+1}^c + h_{t+1}^s + h_{t+1}^p$$

$$\left[\begin{array}{l}
L_t(x_t,u_t)+J_{t+1}^*(x_t) = \frac{1}{2}\begin{pmatrix} x_t \\ u_t \end{pmatrix}'W_t\begin{pmatrix} x_t \\ u_t \end{pmatrix} + \begin{pmatrix} x_t \\ u_t \end{pmatrix}'\begin{pmatrix} w_t^x \\ w_t^u \end{pmatrix} + w_t^c \\[4pt]
\quad + \frac{1}{2}x_t'H_{t+1}x_t + x_t'h_{t+1}^x + h_{t+1}^c + h_{t+1}^s + h_{t+1}^p \\[4pt]
= \frac{1}{2}\begin{pmatrix} x_t \\ u_t \end{pmatrix}'\begin{pmatrix} W_t^{xx} & W_t^{xu} \\ W_t^{ux} & W_t^{uu} \end{pmatrix}\begin{pmatrix} x_t \\ u_t \end{pmatrix} + \begin{pmatrix} x_t \\ u_t \end{pmatrix}'\begin{pmatrix} k_t^x - h_{t+1}^x \\ w_t^u \end{pmatrix} \\[4pt]
\quad + w_t^c + \frac{1}{2}x_t'(K_t - W_t^{xx})x_t + x_t'h_{t+1}^x + h_{t+1}^c + h_{t+1}^s + h_{t+1}^p \\[4pt]
= \frac{1}{2}x_t'W_t^{xx}x_t + \frac{1}{2}x_t'W_t^{xu}u_t + \frac{1}{2}u_t'W_t^{ux}x_t + \frac{1}{2}u_t'W_t^{uu}u_t + x_t'k_t^x \\[4pt]
\quad -x_t'h_{t+1}^x + u_t'w_t^u + \frac{1}{2}x_t'K_tx_t - \frac{1}{2}x_t'W_t^{xx}x_t + x_t'h_{t+1}^x + w_t^c + h_{t+1}^c \\[4pt]
\quad + h_{t+1}^s + h_{t+1}^p \\[4pt]
= (\frac{1}{2}x_t'K_tx_t + \frac{1}{2}x_t'W_t^{xu}u_t + \frac{1}{2}u_t'W_t^{ux}x_t + \frac{1}{2}u_t'W_t^{uu}u_t) \\[4pt]
\quad + (x_t'k_t^x + u_t'w_t^u) + w_t^c + h_{t+1}^c + h_{t+1}^s + h_{t+1}^p \\[4pt]
= \frac{1}{2}\begin{pmatrix} x_t \\ u_t \end{pmatrix}'\begin{pmatrix} K_t & W_t^{xu} \\ W_t^{ux} & W_t^{uu} \end{pmatrix}\begin{pmatrix} x_t \\ u_t \end{pmatrix} + \begin{pmatrix} x_t \\ u_t \end{pmatrix}'\begin{pmatrix} k_t^x \\ w_t^u \end{pmatrix} \\[4pt]
\quad + w_t^c + h_{t+1}^c + h_{t+1}^s + h_{t+1}^p
\end{array}\right]$$

Replacing x_t by the linearized system $x_t = A_tx_{t-1}+B_tu_t+c_t+\xi_t$ we get after the collection of terms:

$$L_t(x_t,u_t)+J_{t+1}^*(x_t) = \tfrac{1}{2}x_{t-1}'A_t'K_tA_tx_{t-1} + \tfrac{1}{2}x_{t-1}'[B_t'K_tA_t + W_t^{ux}A_t]'u_t$$

$$+\tfrac{1}{2}u_t'[B_t'K_tA_t + W_t^{ux}A_t]x_{t-1}$$

$$+\tfrac{1}{2}u_t'[B_t'K_tB_t + 2W_t^{ux}B_t + W_t^{uu}]u_t$$

$$+x_{t-1}'[A_t'K_tc_t + A_t'K_t\xi_t + A_t'k_t^x]$$

7 APPENDIX

$$+u'_t[B'_tK_tc_t + B'_tK_t\xi_t + B'_tk^x_t + W^{ux}_tc_t + W^{ux}_t\xi_t + w^u_t]$$

$$+\tfrac{1}{2}c'_tK_tc_t + c'_tK_t\xi_t + \tfrac{1}{2}\xi'_tK_t\xi_t + c'_tk^x_t + \xi'_tk^x_t$$

$$+w^c_t + h^c_{t+1} + h^s_{t+1} + h^p_{t+1}$$

Next we can calculate $J^*_t(x_{t-1})$ as

$$J^*_t(x_{t-1}) = E_{t-1}(L_t(x_t,u_t) + J^*_{t+1}(x_t))$$

We assume that H_{t+1}, h^x_{t+1}, h^c_{t+1}, h^s_{t+1} and h^p_{t+1} are non-stochastic or known after x_{t-1} has been realized. Then using the expectations (16), (17) and (31) and the analogous expressions listed following (37) in section 3.3, together with (44)-(45), we can see that

$$J_t(x_{t-1},u_t) = \tfrac{1}{2}\begin{pmatrix}x_{t-1}\\u_t\end{pmatrix}'\begin{pmatrix}\Lambda^{xx}_t & \Lambda^{xu}_t\\ \Lambda^{ux}_t & \Lambda^{uu}_t\end{pmatrix}\begin{pmatrix}x_{t-1}\\u_t\end{pmatrix} + \begin{pmatrix}x_{t-1}\\u_t\end{pmatrix}'\begin{pmatrix}\lambda^x_t\\\lambda^u_t\end{pmatrix}$$
$$+\lambda^c_t + \lambda^s_t + \lambda^p_t$$

By minimizing this function the feedback rule yields: $u_t = G_t x_{t-1} + g_t$, where $G_t = -(\Lambda^{uu}_t)^{-1}\Lambda^{ux}_t$ and $g_t = -(\Lambda^{uu}_t)^{-1}\lambda^u_t$. We set this feedback rule into $J_t(x_{t-1},u_t)$ and can derive $J_t(x_{t-1})$:

$$J^*_t(x_{t-1}) = \tfrac{1}{2}x'_{t-1}H_tx_{t-1} + x'_{t-1}h^x_t + h^c_t + h^s_t + h^p_t$$

So it is true for all $t = 1, ..., T+1$.

7.3 Software description

The OPTCON2 algorithm allows the user to find the open-loop as well as the open-loop feedback (passive learning) solutions to optimum control problems for nonlinear (and linear) econometric models with additive and multiplicative uncertainties. The algorithm is implemented in the computer language C# and the corresponding software is named OPTCON 2.0. This software is a useful tool for

- solving stochastic control problems using the OL strategy
- solving stochastic control problems using the OLF strategy
- solving deterministic control problems using the OL strategy
- solving partial stochastic control problems using the OL strategy[29]
- solving partial stochastic control problems using the OLF strategy.

The part of the code with the passive learning strategy is implemented by using the idea of the Kalman filter for updating parameter estimates and is embedded in a Monte Carlo simulation framework.

Interface

The OPTCON 2.0 software provides a Windows-style interface aiming at the user-friendly application of the OPTCON2 algorithm.

Due to the extremely large amount of programming code, it is not possible to include it in this book in print.[30] However a brief overview should be given here. Thus, the usage of the OPTCON 2.0 software, namely the body structure of the main window and the functions of the buttons are described next. When performing an experiment the buttons at the top of the main window are to be clicked in the

---
[29]Partial stochastic control problems are problems where only part of the parameters are stochastic, but the remaining parameters are constant.
[30]The code of OPTCON 2.0 is available on request.

7 APPENDIX

following order (from left to right on the main window): Model, Data, Methodology, Run and Results. Let us discuss each button in depth.

1. Model

   The 'Model' button serves for choosing the model which is implicitly included in C#. At the moment it is possible to choose between the following models: SLOVNL, SLOVL, MacRae, Abel, Hall-Taylor, SLOPOL4 and SLOPOL8. Running the algorithm with these models can be used for recalculating the results presented by the authors or for rerunning these models in a changed mode depending on the questions the user is interested in. Moreover it can be also used as a teaching instrument to learn about the research field of optimal control of nonlinear dynamic models.

2. Data

   This button opens the dialog form for importing the data from excel files. The user is able to name the corresponding files with the appropriate values of state, control, exogenous and other variables. Moreover, such information like the number of Monte Carlo runs, discount factor α, the start and the terminal time periods of the planning horizon, convergence criterion for the system solver etc. can be chosen within this button.

3. Methodology

   This button has two functions:

   - to specify the model type, namely whether the model is deterministic, stochastic or partially stochastic

   - to choose the kind of control strategy: open-loop or open-loop feedback

4. Run

   Clicking on this button starts the algorithm.

5. Results

   The user gets the results in a window that appears after running the program. Using the button 'results' the user can export the results to Excel.

Two additional buttons 'OPTCON' and 'Info' should be mentioned. The 'OPTCON' button is still under construction and does not contain any functions yet except the button which closes the application. One the planned function is choosing a custom working directory. The 'Info' button has to deliver information or help about the OPTCON 2.0 software and is still being developed.

Finally, some advantages and disadvantages of the OPTCON 2.0 software should be mentioned. Among the disadvantages of the OPTCON 2.0 software the following characteristics can be counted:

1. It is not yet possible to input and to modify the structure of a model via the interface. The modification means changes such as altering the set of variables and equations which are in a model. The actions have to be done in the programming code.

2. Choosing a custom working directory is not possible in the present version of the software. All working files have to be located in the path 'C/OPTCON'. It is desirable to modify the software in such a way that this function is available in the next version.

7 APPENDIX

The OPTCON 2.0 software has the following advantageous features:[31]

1. Modern computer language *C#*.

2. The user does not have to be an expert in the field of stochastic control problems (and the different control strategies) in order to make effective use of the software.

3. It is possible to save a set of solutions (by using several runs of the Monte Carlo simulation) as an Excel file and therefore to produce a graph of the results as desired. This feature opens a wide 'spectrum' for analyzing the results and for a better comparison of different control strategies.

---

[31] An ('a priori') advantage of the OPTCON 2.0 software is the fact that this software delivers an optimal solution to the special kind of optimum control problems as described in Section 2.

7 APPENDIX

7.4 Diagrams

The results of the SLOVNL model

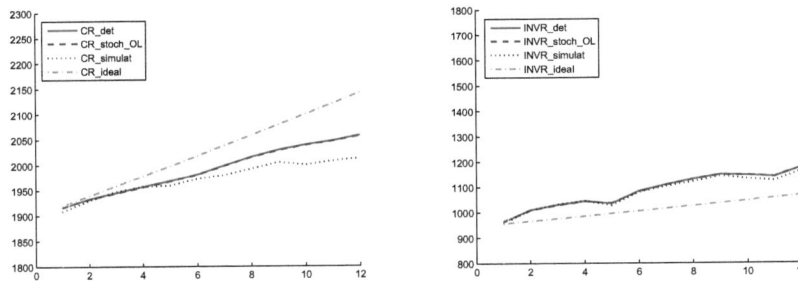

Figure 16: SLOVNL: det. vs. stoch. OL case (CR and INVR)

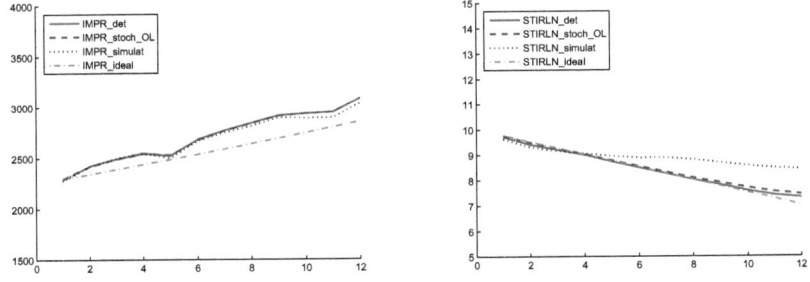

Figure 17: SLOVNL: det. vs. stoch. OL case (IMPR and STIRLN)

7 APPENDIX

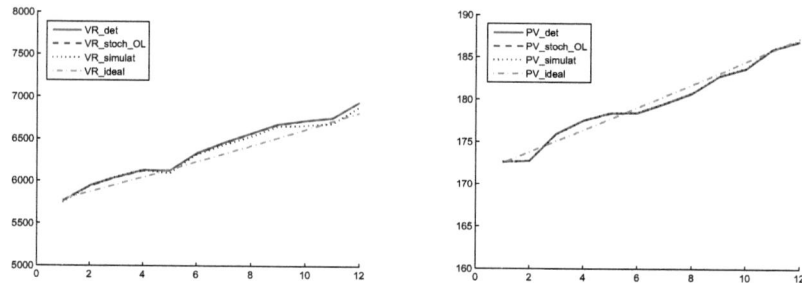

Figure 18: SLOVNL: det. vs. stoch. OL case (VR and PV)

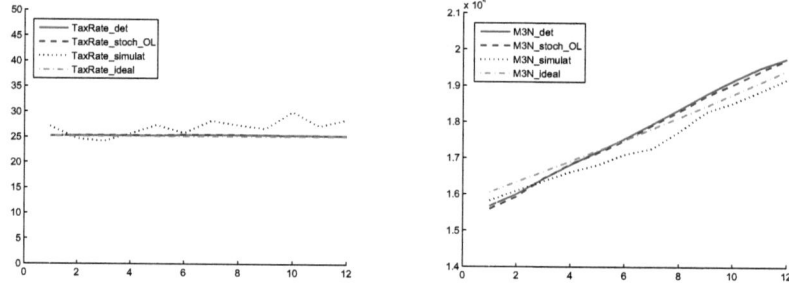

Figure 19: SLOVNL: det. vs. stoch. OL case (TaxRate and M3N)

7 APPENDIX

The results of the SLOVL model

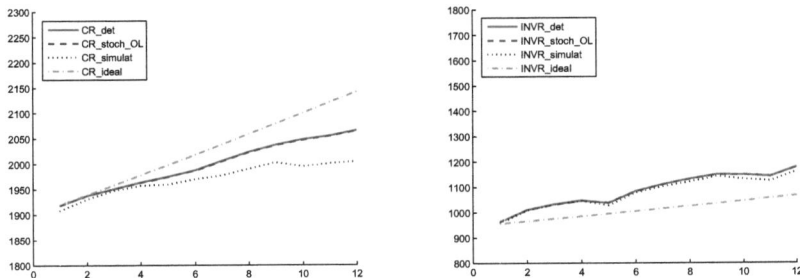

Figure 20: SLOVL: det. vs. stoch. OL case (CR and INVR)

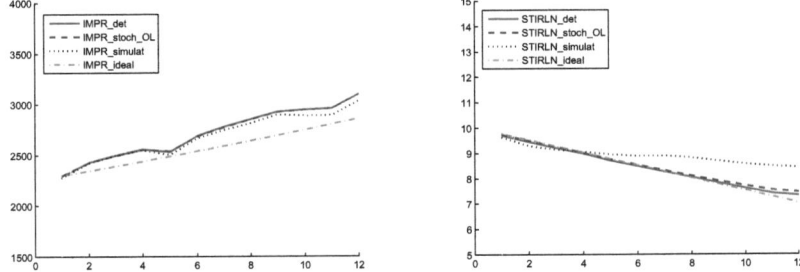

Figure 21: SLOVL: det. vs. stoch. OL case (IMPR and STIRLN)

7 APPENDIX

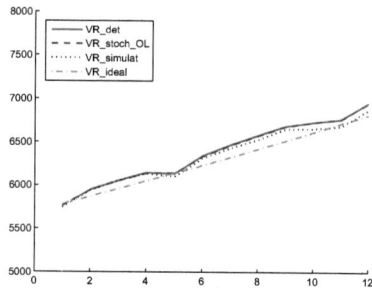

Figure 22: SLOVL: det. vs. stoch. OL case (VR)

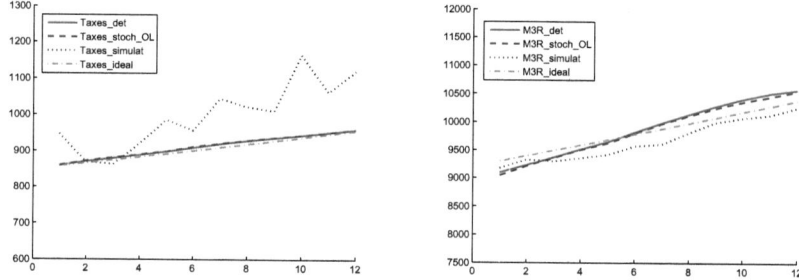

Figure 23: SLOVL: det. vs. stoch. OL case (Taxes and M3R)

8 Acknowledgements

I would like to express my thanks to Prof. R. Neck for his support and expert guidance, and to D. Blueschke for the successful and fair collaboration.

References

Abel, A. B., 1975. A comparison of three control algorithms to the monetarist-fiscalist debate. Ann. Econ. Soc. Meas. Volume 4 (Number 2), 239–252.

Amman, H. M., 1996. Numerical methods for linear-quadratic models. In: Amman, H. M., Kendrick, D. A., Rust, J. (Eds.), Handbook of Computational Economics. Vol. I. Elsevier, Amsterdam, Ch. 13, pp. 587–618.

Amman, H. M., Kendrick, D. A., 1999. Should macroeconomic policy makers consider parameter covariances? Computational Economics 14, 263–267.

Aoki, M., 1967. Optimization of Stochastic Systems. Academic Press, New York.

Aoki, M., 1989. Optimization of Stochastic Systems. Topics in Discrete-Time Dynamics, 2nd Edition. Academic Press, New York.

Arrow, K. J., 1968. Applications of control theory to economic growth. American Mathematical Society 12.

Athans, M., Falb, P. L., 1966. Optimal Control. McGraw-Hill, New York.

Bar-Shalom, Y., Sivan, R., 1969. On the optimal control of discrete-time linear systems with random parameters. IEEE Trans. Autom. Control 14, 3–8.

Bar-Shalom, Y., Tse, E., 1976a. Caution, probing and the values of information in the control of uncertain systems. Ann. Econ. Soc. Meas. 5(2), 323–338.

Bar-Shalom, Y., Tse, E., 1976b. Concepts and methods in stochastic control. Control Dynam. Syst.: Adv. Theory Appl. 12, 99–172.

Bellman, R., 1957. Dynamic Programming, 3rd Edition. Princeton University Press, Princeton, NJ.

Blueschke, D., Blueschke-Nikolaeva, V., Neck, R., 2012. Stochastic control of linear and nonlinear econometric models: Some computational aspects. Computational Economics.

Blueschke-Nikolaeva, V., Blueschke, D., Neck, R., 2012. Optimal control of nonlinear dynamic econometric models: An algorithm and an application. Computational Statistics and Data Analysis 56 (11), 3230–3240.

Bryson, A. E., Ho, Y.-C., 1969. Applied Optimal Control. Blaisdell, Walham, Mass.

Bryson, A. E., Ho, Y.-C., 1975. Applied Optimal Control. Optimization, Estimation, and Control. Hemisphere, Washington, DC.

Chen, B., Zadrozny, P. A., 2009. Multi-step perturbation solution of nonlinear differentiable equations applied to an econometric analysis of productivity. Computational Statistics and Data Analysis 53, 2061–2074.

Chow, G. C., 1967. Multiplier, accelerator and liquidity preference in the determination of national income in the United States. Review of Economics and Statistics 49.

Chow, G. C., 1973. Effect of uncertainty on optimal control policies. Int. Econ. Rev. 14.

Chow, G. C., 1975. Analysis and Control of Dynamic Economic Systems. John Wiley & Sons, New York.

REFERENCES

Chow, G. C., 1981. Econometric Analysis by Control Methods. John Wiley & Sons, New York.

Chow, G. C., Butters, E. H., 1977. Optimal control of nonlinear systems program: user's guide. Tech. rep., Econometric Research Program, Princeton University.

Coomes, P. A., 1987. PLEM: A computer program for passive learning, stochastic control experiments. Journal of Economic Dynamics and Control 11, 223–227.

Curry, R. E., 1969. A new algorithm for suboptimal stochastic control. IEEE Trans. Autom. Control AC-14, 533–536.

Friedman, B. M., Howrey, E., 1973. Nonlinear models and linear optimal policies, discuss. paper 316, Harvard Inst. Econ. Res.

Holt, C. C., 1962. Linear decision rules for economic stabilization and growth. Econometrica 76, 20–45.

Kalman, R. E., 1960. A new approach to linear filtering and prediction problems. Journal of Basic Engineering 82D, 33–45.

Kendrick, D. A., 1973. Stochastic control in macroeconomic models. Inst. Elec. Eng. IEEE Conf. publ. 101, 200–207.

Kendrick, D. A., 1981. Stochastic Control for Economic Models. McGraw-Hill, New York, second edition 2002 online at www.eco.utexas.edu/faculty/Kendrick.

Kendrick, D. A., Amman, H. M., 2006. A classification system for economic stochastic control models. Computational Economics 27, 453–481.

Kendrick, D. A., Coomes, P., 1984. Dual, a program for quadratic-linear stochastic control problems. Center for Economic Research, University of Texas, Austin, Discussion paper 84–15.

Kendrick, D. A., Taylor, L., 1970. Numerical solutions of nonlinear planning models. Econometrica 38, 453–467.

Livesey, D., 1971. Optimizing short-term economic policy. Econ. J 81, 525–546.

MacRae, E. C., 1972. Linear desicion with experimentation. Ann. Econ. Soc. Meas. Vol. 1 (No. 437-447).

MacRae, E. C., September-November 1975. An adaptive learning rule for multiperiod decision problems. Econometrica Vol. 43 (No. 5-6).

Matulka, J., Neck, R., 1992. OPTCON: An algorithm for the optimal control of nonlinear stochastic models. Annals of Operations Research 37, 375–401.

Neck, R., 1984. Stochastic control theory and operational research. European Journal of Operations Research 17, 283–301.

Neck, R., Blueschke, D., Weyerstrass, K., 2011. Optimal macroeconomic policies in a financial and economic crisis: a case study for Slovenia. Empirica 38, 435–459.

Neck, R., Haber, G., Weyerstrass, K., 2010. Optimal deterministic and stochastic macroeconomic policies for Slovenia: An application of the optcon algorithm. Computation Economics 36, 37–45.

Neck, R., Karbuz, S., 1995. Optimal budgetary and monetary policies under uncertainty: a stochastic control approach. Annals of Operations Research 58, 379–402.

REFERENCES

Neck, R., Karbuz, S., 1997. Optimal control of fiscal policies for Austria: Applications of a stochastic control algorithm. Nonlinear Analysis, Theory, Methods and Applications 30, 1051–1061.

Neck, R., Karbuz, S., 2000. On the influence of stochastic parameters on optimal macroeconomic policies. Computation in economic, finance and engineering: Economic systems 58, pp. 423–428.

Neck, R., Matulka, J., 1992. Optcon: An algorithm for the optimal control of nonlinear stochastic models. Annals of Operations Research 37, 375–401.

Neck, R., Matulka, J., 1994. Stochastic control of nonlinear economic models. New Directions in Computational Economics, Advances in Computational Economics 4, 207–226.

Norman, A. L., 1976. First order dual control. Ann. Econ. Soc. Meas. 5 (3), 311–322.

Phillips, A. W., 1954. Stabilization policy in a closed economy. Econ.J. 64, 290–323.

Phillips, A. W., 1957. Stabilization policy and the time form of the lagged responses. Econ.J. 67, 265–277.

Pitchford, J., Turnovsky, S. J., 1977. Application of Control Theory to Economic Analysis. North-Holland, Amsterdam.

Samimi, A. J., Tehranchian, A. M., 2005. An application of the stochastic optimal control. Algorithm (OPTCON) to the public sector economy of Iran. Iranian Economic Review 10(13).

Samimi, A. J., Tehranchian, A. M., Rad, M. A., 2010. Optimal combinations of government expenditures to economic growth process in Iran. Journal of Applied Sciences Research 6(5), 387–392.

Shupp, F. R., 1972. Uncertainty and stabilization policies for a nonlinear macroeconomic model. Q. J. Econ 80, 94–110.

Tehranchian, A. M., Poorhabib, R., Karegar, N., Behravesh, M., 2011. Dynamic optimization of LQ objective loss function: Application in economic planning. Journal of American Science 7(5), 656–660.

Theil, H., 1957. A note on certainty equivalence in dynamic planning. Econometrica 25, 346–349.

Tse, E., Athans, M., 1972. Adaptive stochastic control for a class of linear systems. IEEE Trans. Autom. Control 17, 38–52.

Tustin, A., 1953. The Mechanism of Economic Systems. Harvard University Press, Cambridge.

Welch, G., Bishop, G., 2006. An introduction to the kalman filter. Tech. rep., Department of Computer Science, University of North Carolina at Chapel Hill.

Weyerstrass, K., 1999. Optimal monetary and fiscal policy for Slovenia under different exchange rate regimes. Tech. rep., University of Klagenfurt, Department of Economics.

Weyerstrass, K., Haber, G., Neck, R., 2001. SLOPOL1: A macroeconomic model for Slovenia. International Advances in Economic Research (7(1)), 20–37.

Weyerstrass, K., Neck, R., 2007. SLOPOL6: A macroeconometric model for Slovenia. International Business and Economics Research Journal (6(11)), 81–94.

i want morebooks!

Buy your books fast and straightforward online - at one of world's fastest growing online book stores! Environmentally sound due to Print-on-Demand technologies.

Buy your books online at
www.get-morebooks.com

Kaufen Sie Ihre Bücher schnell und unkompliziert online – auf einer der am schnellsten wachsenden Buchhandelsplattformen weltweit! Dank Print-On-Demand umwelt- und ressourcenschonend produziert.

Bücher schneller online kaufen
www.morebooks.de

VDM Verlagsservicegesellschaft mbH
Heinrich-Böcking-Str. 6-8   Telefon: +49 681 3720 174   info@vdm-vsg.de
D - 66121 Saarbrücken   Telefax: +49 681 3720 1749   www.vdm-vsg.de

Printed by Books on Demand GmbH, Norderstedt / Germany